RIGHT TO MANAGE AND SERVICE CHARGES: THE NEW REGIME

by

Brian Jones

A division of Reed Business Information

Estates Gazette, 151 Wardour Street, London W1F 8BN

ISBN 0 7282 0428 2

Typeset by Amy Boyle, Rochester, Kent
Printed in Great Britain by Bell & Bain Ltd., Glasgow

Contents

Table of Statutes

Table of Statutory Instruments

Preface

The Commonhold and Leasehold Reform Act 2002 received Royal Assent on May Day 2002. Apart from the commencement of a handful of sections on 26 July 2002, the vast majority of the Act's provisions did not come into force until well over a year after its passage. Indeed, much of it still remains to be implemented, especially in Wales. For example, the Commencement Orders of September and October 2003 related only to England.

Never has it been truer that 'the devil is in the detail'. In order to implement the Act it has been necessary to create an enormous body of secondary legislation. Consultation has been an exhaustive and exhausting process. Meanwhile, all of us involved in this field have had to speculate about what the final fully-fledged version of the Act would mean in practice. Conferences and seminars have abounded, the property press has been full of articles and correspondence, entire training programmes have been built around the proposals, and marketeers have been busy assessing potential opportunities. Given the length of its gestation period, and the additional uncertainty which was created when the Bill was dropped to make way for the 2001 General Election, there has probably been more rumination about this Act than any other statute affecting the residential leasehold sector in living memory.

At last, we are in a position to look at the end result.

The first part of the Act introduces Commonhold – a new, third form of tenure in land. As can easily be surmised from the government's consultation paper of August 2000 (which received no serious opposition to its principle) it is Parliament's intention that Commonhold will eventually replace leasehold as a means of ownership of residential units in multiple developments. Leaving aside the question of whether that is an optimistic appraisal, the process will take a very long time: generations rather than years perhaps. Consequently, the second part of the Act sets out yet further changes to the law of leasehold, in order to impose additional restraints upon landlords and instigate new rights for long leasehold tenants.

The purpose of this book is to examine the leasehold reform elements of the Act, taking a practical approach to the immediate and foreseeable effects upon the management of blocks of flats. I hope that it will prove to be of valuable assistance to professionals and lay people alike, whether they are or act for landlords or tenants.

The aim is to concentrate on issues which are likely to arise in practice and give some guidance on how they may be dealt with. It is not my intention to produce a legalistic commentary on the Act. Neither have I set out to cover commonhold nor the non-contentious areas of leasehold enfranchisement in any detail. Those aspects have been explained by others, better qualified to do so. In any event, the Commonhold provisions are not expected to be implemented until the Spring of 2004 at the earliest.

I have spent the majority of my professional life endeavouring to assist landlords, tenants and managers of blocks of flats in coping with the obstacles thrown at them by legislation, the courts and, in more recent times, the Leasehold Valuation Tribunals. I have conducted cases large and small across the country and across the perceived landlord/tenant divide. I believe that my experience helps me to identify the most common problems which have occurred, and thus those likeliest to crop up under the new regimes introduced by this Act.

This book is divided into four principal sections.

Part 1 – The Right to Manage

This is an entirely new concept: that long leasehold tenants should be able to take over the management of their properties without either acquiring the freehold or having to prove fault on the part of the landlord or the current managers.

Right to Manage ('RTM') will bring about profound changes for many leaseholders – and indeed landlords and property managers. The procedures to achieve the right are superficially straightforward, but there are many complexities and difficulties just under the surface.

The exercise of RTM will usher in a whole host of dangers for landlords and tenants alike if it is not managed efficiently. Threats are present as well as opportunities. I will be looking at both the positive and negative sides, and suggesting preparatory measures towards a successful implementation of the new rights.

Part 2 – Service charges

The legislation governing residential service charges last received a major overhaul in the Landlord and Tenant Acts of 1985 and 1987. The 2002 Act does not replace them, but puts through many significant re-writes.

The new Act will alter fundamentally:

- Service charge accounting.
- The consultation processes.
- The dispute resolution mechanisms, including jurisdiction as between the courts and the Leasehold Valuation Tribunals.
- Arrears recoveries.
- The way insurance is dealt with.
- The grounds for obtaining variations of leases in the service charge context.
- The definition of service charge itself.

The changes to the law on service charges will have far-reaching consequences on any party responsible for accounting for and collecting leaseholders' monies. Some of those consequences will lead to draconian penalties in the event of non-compliance.

I will be highlighting and explaining the changes, and indicating ways in which penalties can be avoided.

Part 3 – Ground rent and forfeiture

In contrast to service charges, the law relating to ground rent has remained untouched by statute in any substantive sense since Victorian times. This is no longer the case. The changes in the 2002 Act affecting ground rent are fundamental if brief.

Forfeiture, however, is a subject which has been tackled by Parliament on numerous occasions – most recently the Housing Act 1996. Although Parliament has still drawn back from outright abolition, this Act brings in additional restraints and restrictions against forfeiture of residential leases. The prospects of a successful forfeiture will recede almost to nothing, and accordingly the threat of it will be much diminished.

For some time now, it has been the threat rather than the fact of forfeiture which has provided a valuable weapon to landlords (of whatever type). As the weapon loses its sharpness, landlords will have to look increasingly to other methods of enforcement of covenants.

I will be examining how forfeiture may still be used at all and at the alternatives available for various breaches of covenant. I will also be looking at the new role of the Leasehold Valuation Tribunals in the enforcement context.

Part 4 – Conclusions

Every novel must have an ending, and we are looking here at some very novel ideas. This Act however is far from the last word on reforming leasehold law. While leases exist (and, despite the creation of commonhold, there is little to suggest that leases have yet reached the end of the road) there will always be new legislation in the pipeline. For example, the Law Commission is still carrying out its seemingly lifelong mission to find a replacement for forfeiture which is acceptable to all sides.

It is sometimes forgotten that there are no longer just landlords and tenants involved in leasehold flats; indeed that ceased to be the case a long time ago. It has been estimated that there are currently around 1.5 million long leasehold flats in the private sector, and that perhaps 30% of them are held by members of companies which own their own freeholds.

The definitions of 'landlord' and 'tenant' therefore have become extremely vague. Of course, there is also a large variety of professional advisers to the parties. A short list of those who will feel the effects of the Act includes:

- Developers
- Institutional landlords
- Local authorities and Housing Associations
- Leaseholders
- Property managers
- Surveyors
- Valuers
- Arbitrators
- Solicitors and other lawyers
- Residents' Management Companies
- Right to Enfranchise companies
- Right to Manage companies

I could go on.

I shall attempt to summarise the most likely effects of the 2002 Act on the wide variety of parties concerned with blocks of flats and try to identify the principal areas in which steps can be taken to minimise the Act's most disruptive consequences.

Brian Jones
November 2003

PART 1

RIGHT TO MANAGE

Chapter 1

Qualification and the Preliminary Steps

Qualification for Right to Manage (RTM) is superficially straightforward.

Qualifying tenants

A qualifying tenant, subject to a few limited exceptions, is any long leaseholder of a flat (section 75(2)). The exceptions are set out in section 75(3)–(6) Commonhold and Leasehold Reform Act 2002 ('CLRA'). They are:

- Business tenancies.
- A lease granted out of a superior lease other than a long lease, in breach of the terms of the superior lease where the breach has not been waived by the superior landlord.
- There can be no more than one qualifying tenant for a flat at any point in time.
- Where there is more than one lease, any superior lease will not qualify.

In the case of joint tenants, they will be regarded as a single qualifying tenant (section 75(7)). As we shall see later, there are certain consequential problems arising from this point when it comes to voting rights and assessing the numbers for qualification for RTM. Similar complexities will occur in relation to tenants in common, who are not mentioned in this context in the Act.

A 'long lease' is defined by sections 76 and 77 of the Act. Essentially, it is a long lease if:

- It is for a term exceeding 21 years, whether or not it is terminable by notice by either side or by forfeiture within that period.
- It is for a term fixed by law with a covenant or obligation for perpetual renewal.
- It is a lease terminable after death or marriage (section 149(6) Law of Property Act 1925).

- It was acquired under the 'right to buy' legislation (Housing Acts 1985 and 1996).
- It is a shared ownership lease where the tenant's total share is 100%.
- The lease was originally for a term of less than 21 years, but has since been renewed (without payment of a premium) for a term totalling in excess of 21 years, or
- There is more than one lease for a flat between the same parties, and the flat is wholly demised by reading the leases together.

Thus, taking a simple lease structure, the tenant of a residential flat under a lease for more than 21 years will qualify to participate in RTM, whether or not he* lives at the property and no matter how long he has owned the flat.

Qualifying premises

The next question to consider is what premises qualify. The criteria here are set out in section 72 of the Act. Again, the criteria are notionally straightforward. Premises will qualify for RTM if:

- They consist of a self-contained (meaning structurally detached) building or part of a building, with or without appurtenant property.
- Qualifying tenants hold two or more flats.
- The total flats held by qualifying tenants make up at least two-thirds of the flats in the building.
- They are not exempt by virtue of Schedule 6 to the Act.

The rules are a little more complicated in the case of a part of a building. Section 72(4) states that it will be regarded as self-contained if it constitutes a vertical division of the building which could be redeveloped independently of the remainder of the building and services are provided independently or could be so provided (without a 'significant interruption' to the services to other occupiers).

Schedule 6 provides that premises are excepted if:

- 25% of the internal floor area is used for non-residential purposes.

---

* The pronoun 'he' is used generically throughout the book, to include both he and she.

- The freehold of different self-contained parts of the premises is in different ownership.
- There are no more than four units in a property converted into flats and the freeholder (or an adult member of his family) has occupied a qualifying flat for the last 12 months (see paragraph 3 of Schedule 6 for further details).
- The immediate landlord of any of the qualifying tenants is a local housing authority.
- Premises where RTM has already been exercised within the last four years (unless the freehold had been acquired by the RTM company so that it had ceased to be an RTM company).

It is worth noting at this stage that there is no bar to RTM where the freehold is already held (or the property already managed) by a Residents' Management Company, a Right to Enfranchise company or the like.

Put simply then, a self-contained building or part of a building containing two or more flats let on long leases and which does not contain a significant commercial area is likely to qualify for RTM. In the case of an estate consisting of more than one building, RTM could be exercised separately in respect of each building which qualifies in its own right.

The Right to Manage company

Having established that the premises qualify and that there are sufficient qualifying tenants, there is only one remaining requirement before Right to Manage can be exercised. This is not an obligation to prove fault on the part of the existing manager (as is pointed out in the Introduction), but the necessity to have formed a Right to Manage company which complies with sections 73 and 74 of the Act.

A RTM company must:

- Be a private company limited by guarantee (as opposed to shares). The requirement is for a guarantee of a sum not exceeding £1 from each member.
- State in its memorandum of association that its object (or one of them) is to acquire and exercise the right to manage the premises concerned.

But it cannot be an RTM company if:

- It is a Commonhold Association (under Part 1 of the Act).

- There is already an RTM company in respect of the same premises.
- It has acquired the freehold of the premises.

Section 74(1) provides that each qualifying tenant is entitled to be a member of the RTM company. From the date that the right to manage is eventually acquired ('the acquisition date' – see below), any landlords under any leases for the whole or any part of the premises (including, obviously the freeholder) are also entitled to membership (section 74(1)(b)).

So far as the formalities for the company are concerned, section 74(7) of the Act exempts it from sections 2(7), 3 and 8 of the Companies Act 1985, so as not to fetter the uniqueness of an RTM company as a creature of the 2002 Act. Otherwise, the regulations made by secondary legislation under the 2002 Act[1] prescribe the content and form of the company's governing instrument, its memorandum and articles of association. In the event that an RTM company was formed before 30 September 2003 (when the regulations came into force), its memorandum and articles will be deemed to incorporate the prescribed contents to secure compliance with both the Act and the regulations.

Clause 4 of the prescribed memorandum covers the powers the company may use in furtherance of its objects, and in particularising them provides a useful checklist for the range of things an RTM company will have to be ready to do from acquisition of the Right to Manage. Perusal of this clause is advisable for RTM company directors in this context, quite apart from gaining an understanding of what is required by the memorandum and articles generally.

The most startlingly different topic covered in the memorandum and articles of association of an RTM company is how voting rights are to work, particularly after the acquisition date when the landlord is entitled to join.

A detailed analysis of the practical effects of the voting system, and the decision-making processes generally, can be found at Chapter 7.

Preliminary considerations

By drawing together section 79 (4) and (5), it can be seen that the very minimum number who can exercise the Right to Manage in

---

[1] The RTM Companies (Memorandum and Articles of Association) Regulations 2003.

the very smallest block is two qualifying tenants. For practical purposes, two is also the minimum number for setting up an RTM company and proceeding to the next preliminary step: the service of the Notice Inviting Participation.

It is at this stage that any leaseholders should be giving serious consideration to obtaining professional advice and assistance, if they have not done so already. It should be possible to set up the RTM company without help as the documentation and guidance notes will be easily accessible from Companies House. Further free assistance will be available from the likes of the Leasehold Advisory Service ('LEASE').

Setting up a company is one thing; managing a block of flats quite another. As we shall see from consideration of other parts of the Act, and as is evident from earlier statutes in any event, managing residential developments is fraught with legal and practical complexities which far exceed those concerned with practically any other endeavour: for example, Landlord and Tenant Acts 1985 and 1987.

Among the points which need to be examined at this very early stage (and upon which advice should be sought) are:

- Is Right to Manage the best way to achieve the result sought?
- Always consider alternatives such as purchasing the freehold (perhaps using the threat of RTM as a bargaining chip) or applying to the Leasehold Valuation Tribunal for the appointment of a manager.
- Do the leaseholders have the time, energy and expertise between them to take on the task?
- Are the leases sufficiently well drafted to allow an RTM company to function efficiently?
- Will adequate funding be available to keep up repair and maintenance programmes?
- Has a business plan been prepared?
- Is adequate liability insurance available?
- Are there any special features affecting the building which will exacerbate the difficulties of management (for example, a high proportion of sub-let flats)?

Professionals who ought to be in a position to give helpful advice include:

- Managing Agents (details can be obtained from organisations such as ARMA (the Association of Residential Managing Agents), ARHM (the Association of Retirement Housing

Managers) and LEASE.
- Solicitors and other lawyers.
- Surveyors and valuers.
- Accountants.
- Some local authorities operate Housing Advice Centres, who may be able to assist.

It can generally be assumed that managing agents of residential properties will have the necessary expertise, and indeed may well have links to other professionals who are appropriately equipped. With others such as solicitors, surveyors and accountants there may only be a few in a given location (or none at all in some places) who have developed relevant expertise in this field (which is not always a tremendously popular practice area). It is advisable to ask searching questions concerning their abilities, or at least make inquiries of independent bodies such as LEASE, ARMA or ARHM before committing to a particular firm.

Notice inviting participation

Assuming that the intention is still to proceed, the RTM company must prepare the Notice Inviting Participation under section 78. This is a mandatory procedure, although presumably it could be dispensed with if 100% of the qualifying tenants are already on board. Having said that, there may be a risk (particularly in larger developments) that a flat changes hands at around the same time the notice should have been served, and failure to serve the incoming purchaser could invalidate the process. Therefore, it may be prudent to serve it in any event.

Section 78(1) requires that the notice must be served upon each qualifying tenant who has not already joined or agreed to join the RTM company. The notice must follow the prescribed form and content, covering such things as:

- A statement that the company intends to acquire the right to manage the premises.
- The names of the RTM company members.
- An invitation to join the company.
- The company's registered number, its registered office address and the names of its directors and secretary.
- The names of the landlord and any third parties to the leases (such as a residents' management company ('RMC'), presumably whether or not it exists in practice).

- A statement that the company will be responsible for the landlord's duties and powers under the leases (except as provided below) regarding services, repairs, maintenance, improvements, insurance and management.
- A statement that the company may enforce untransferred tenants' covenants.
- A statement that the company will *not* be responsible for the landlord's duties or powers in respect of flats or other units not held by qualifying tenants, nor in respect of re-entry or forfeiture.
- A statement that the company will be subject to the statutory provisions contained in Schedule 7 to the Act (see later).
- A statement as to whether the company intends to appoint a managing agent and, if so, the proposed agent's name and address and if it is the landlord's agent. If not, the notice must set out any qualifications or experience on the part of company members relating to the management of residential property.
- Acknowledgement that company members may be personally liable for the costs of the landlord 'and others' in relation to a subsequent claim notice. ('Others' are not defined. If it was intended that this should be restricted to third parties to the leases, this would be so stated in the legislation (specifically paragraph 3(h) of the Right to Manage (Prescribed Particulars and Forms) (England) Regulations 2003). That it does not do so perhaps implies that 'others' is to include the likes of managing agents.)
- A statement that, if the recipient does not fully understand the purpose or implications of the notice, he should seek professional help.
- The information inserted as notes to the prescribed form. These notes run to some 10 paragraphs, in addition to the 12 paragraphs in the body of the notice and its schedule.

In addition, the notice must be accompanied by a copy of the company's memorandum and articles of association, or details of how and where they may be inspected or copied (section 78(4) and (5)).

Finally, and intriguingly, section 78(7) provides:

> A notice of invitation to participate is not invalidated by any inaccuracy in any of the particulars required by or by virtue of this section.

Note that this does not protect a notice which omits a prescribed item entirely. Neither should this apparently generous provision be taken as licence for deliberate or careless lack of attention to detail. As we shall see later a defective notice will give the landlord an opportunity to delay and possibly circumvent the exercise of RTM, so it is important to prepare the notice as accurately as possible.

Preparation of the section 78 notice will be an onerous and time-consuming task. This underlines the probable need for professional assistance in drawing it up. Likewise, it will not be an easy document to read. Particularly in smaller developments where the numbers of potential members of the RTM company are more manageable, it may be wise to present the notice at a meeting of qualifying tenants when information can be expanded and questions answered. To the disappointment of many however there is no requirement for a business plan to support the notice.

Service of the section 78 notice (as with others in relation to RTM) must be carried out under section 111 of the Act. It must be in writing and *may* be sent by post. The notice may be served at the flat, unless the qualifying tenant has notified the company of an alternative address in England or Wales.

This raises an interesting early problem-solving exercise for the RTM company. It is very unlikely that the qualifying tenants who have not already become members of the company will even know of the RTM company's existence at this stage – especially if they are not resident at the property. How then can they have notified the company of an alternative address?

It may be that an officer of the RTM company has been told informally and in quite another context that the tenant concerned prefers his post to go elsewhere, but there is nothing in the Act to deem such knowledge as notification to the company. Judicial decisions might eventually imply such a deeming provision, but it is likely to be a very long time before such cases come to trial. Similarly, there is no requirement to carry out Land Registry searches; even if there were, HM Land Registry is hardly an accurate guide to the current residential addresses of registered proprietors.

Of course, there could be reasons whereby the RTM directors would be quite content if one or more qualifying tenants did not receive the invitation to join the company. The tenants might be perceived as the landlord under different company names for example, or they might be directors of an existing Residents' Management Company.

All that is required at this stage is that the section 78 notice is served, and that the RTM company ends up with sufficient numbers to proceed to the next step – 50% of qualifying tenants, or both if there are only two (section 79(5)).

Nevertheless, it must be wise to maximise the number of tenants in the company at the very start. Management of a block of flats or similar development is a heavy burden, and building in a potential opposition before RTM has even started would be a dangerous strategy. Even on the simplest level, the costs of getting going (company set-up costs, insurance premiums, the landlord's costs for dealing with RTM) will be significant, and the further they can be spread the better.

Checklist

To summarise the initial steps to exercising RTM:

1. Are there enough qualifying tenants?
2. Do the premises qualify?
3. Set up the RTM company.
4. Obtain professional advice and assistance as necessary.
5. Consider the options.
6. Prepare a business plan (not necessarily required, but advisable).
7. Have funding and insurance available.
8. Prepare and serve the Notice Inviting Participation.

Chapter 2

Notice of Claim

Section 79(2) provides that formal notification of the claim to exercise Right to Manage can be given no earlier than 14 days after the section 78 Notice Inviting Participation has been given to all those entitled to receive it. The Act does not help us to determine whether any additional time should have been allowed for service of the section 78 notice; unless and until judicial guidance is given on this point, it seems likeliest to assume that the 14 days run from the date the notice is hand delivered or put in the post.

Having said that, and given the desirability of maximising numbers, it would be imprudent for the RTM company to rush to serve the claim notice while non-participating qualifying tenants still need a reasonable time to consider their options, unless perhaps there is a special degree of urgency. Even if an emergency is perceived to exist, the remainder of the procedure will still take some months before RTM can be acquired, so the RTM company directors may feel that some flexibility at this stage is warranted.

Of course, if there has been a substantial early take-up of the invitation to participate, there is little point in waiting further.

The claim notice

It is possible to serve the section 79 'notice of claim to acquire the right' ('the claim notice') if the following criteria are met:

- At least 14 days have passed since the Notice Inviting Participation.
- The RTM company's membership must include at least 50% of the qualifying tenants of flats in the premises, or, if there are only two qualifying tenants, both of them (section 79(4) and (5)).

The claim notice must be served upon any and each landlord under a lease for all or part of the premises concerned, any third party to such leases, and any manager who may have been appointed by the court or the LVT under Part 2 of the Landlord and Tenant Act 1987 (section 79(6)). It is necessary to effect service upon the correct parties at the date of the claim notice, so it would be wise to carry out a search against the relevant titles at HM Land Registry a short time before service. Section 79(7) allows for non-service if any of these parties cannot be found or identified (which should not be the case if they hold a registered title), and section 85 deals with the scenario whereby none of the parties can be served (see below).

A copy of the claim notice must also be given to each qualifying tenant (whether or not they are members of the RTM company), thus providing a further check on the RTM company ignoring particular individual leaseholders. If a manager had been appointed in respect of the premises, a copy of the claim notice must be lodged with the court or LVT who appointed him.

Section 111 applies again in relation to service of the claim notice. The notice must be in writing and *may* be sent by post. Specifically in relation to the landlord, notice must be sent to the last address notified to a member of the RTM company under section 48 of the Landlord and Tenant Act 1987 for service of notices or, failing that, the last address notified under section 47 of the 1987 Act in a demand for rent. However, if the landlord has notified the RTM company (presuming he knew of its existence) of a different address in England or Wales precisely for the purpose of service of an RTM claim notice, then that address must be used (section 111(4)).

The last provision also applies to qualifying tenants. It is therefore important to check whether any alternative addresses have been notified since service of the section 78 Notice Inviting Participation.

The claim notice also comes in a prescribed form and with prescribed contents (section 80). The notice must include:

- The name of the premises and a statement of the grounds upon which it is claimed that they qualify for RTM (if the premises form only part of an estate or development under the same freehold ownership, it would be sensible to identify the part which is the subject of the claim).
- A statement of the full names of each qualifying tenant who is a member of the RTM company and the address of his flat. (If any of the names given are not, in fact, qualifying tenants at the date of the notice, the notice will still be valid so long as those who do qualify still amount to at least 50% (section 81(2).)
- Particulars of the lease held by each such person including the date it was entered into, the term and the commencement

date of the term (all of which may be found through HM Land Registry).

- The name and registered office address of the RTM company.
- The period for service of a counternotice: not earlier than one month after the claim notice.
- The acquisition date (see below).
- A statement that a recipient who does not dispute the entitlement to RTM and is the 'manager party' (see section 91(2) and (4)) under a pre-existing 'management contract' (see section 91(2)) must give notice in accordance with section 92 to the 'contractor party' (see section 91(2)(b)) and to the RTM company.
- Notification that any landlords under leases affecting the premises are entitled to become members of the RTM company from the acquisition date.
- A statement that the notice will not be invalidated by any inaccuracy in its particulars, but inviting any recipient who perceives an inaccuracy to identify it with reasons.
- A statement that a recipient who does not fully understand the purpose of the claim notice is advised to seek professional help. (There is a similar statement in the section 78 notice, but in this case it relates only to the purpose of the notice, not its implications.)
- The information inserted as notes to the prescribed form.

The claim notice remains in force until the acquisition date unless it has been withdrawn or is deemed to have been withdrawn or it has otherwise ceased to be of effect (section 81(4)). While the claim notice is in force, no other claim notice may be served affecting the premises (section 81(3)). This last point has some significance, because if the original claim notice is served on behalf of a bare 50% of qualifying tenants, it is conceivable that the other 50% could seek to mount an alternative claim to RTM. Section 81(3) will prevent them from doing so.

The acquisition date

The correct calculation of the acquisition date has significant implications for both the RTM company and the landlord (and indeed the landlord's agents and contractors). It is therefore worth looking at in its own right. Although the date to be inserted in the claim notice is not necessarily decisive (as we shall see later), it is essential that the RTM company works on the assumption that it will be since it is not in its power to change the date.

Put very simply, the acquisition date is the date on which RTM is acquired. From that date, the RTM company becomes responsible for the performance of the landlord's management covenants and, conversely, the landlord ceases to be responsible.

The formula for calculating the date is straightforward enough: from the date of the claim notice the recipients must have at least one month to serve a counternotice, and the acquisition date must be set at least three months after that expires. Accordingly, the acquisition date could be fixed on any date after four months after the claim notice is served. Indeed, if no counternotice is subsequently served, the acquisition date set out in the claim notice will apply.

Obviously it will be tempting to fix the earliest possible date, simply four months from service of the claim notice. However, there are a number of practical considerations to be taken into account, for example:

- Accounting for service charge funds could well be fraught with complexity in any event, but using a year end or at least a quarter day should make things slightly easier.
- Existing contracts will generally terminate on the acquisition date, so it would be prudent to avoid it happening at a sensitive stage in the carrying out of a repairs contract for example.
- Renewal dates for annual contracts such as insurance should be avoided for the same reason.
- The RTM company will be obliged to take over responsibility for matters such as repairs and maintenance immediately on the acquisition date, so it would be unwise to fix it for Christmas Day, Good Friday or some other date when it will be difficult to find a supplier of essential services in a hurry.
- Inevitably, there will be one or two directors of the RTM company who will take the lead. It would not be helpful if the acquisition date coincided with their holidays.

Although four months seems a long time ahead, forward planning is an essential pre-requisite for the exercise of RTM.

Obtaining information

Clearly, the RTM company may need to obtain information from the landlord and any managers of the premises at a variety of stages: in preparing the claim; in preparing to take over management; and in carrying out its management functions after the acquisition date. Accordingly, the Act deals with the right to information at different points, principally sections 82 and 93.

By section 82, an RTM company may give notice to 'any person' requiring him to provide any information which is in that person's possession or control and which the company reasonably requires for the purpose of ascertaining the particulars for the section 79 claim notice. The company may go on to require facilities to inspect records 'in a readily intelligible form' and, for a reasonable fee, copies. The recipient of the notice has 28 days within which to comply (section 82(3)) or faces enforcement through the county court. All of the relevant rights and duties in relation to Right to Manage are enforceable through the county court under section 107. There is no prescribed form for this notice.

Section 82 raises a number of potential problems for landlords, managing agents, and indeed RTM companies. The notice is served specifically to ascertain the particulars for the claim notice, which will largely cover the names and addresses of leaseholders and the particulars of the leases. Apart from their correspondence addresses, this information should be obtainable from HM Land Registry. According to government sources, because this Act post-dates the Data Protection legislation, that will not provide a smokescreen for recalcitrant landlords; but the very request will put the landlord on notice before the event that RTM is envisaged and that not all qualifying tenants are yet on board. Moreover, by this stage, the landlord will have had no opportunity to assess independently the validity of the RTM company, so how is he to know that this is not just a fishing expedition? Such a doubt may give the landlord the chance to be obstructive at least, and drag things out.

There is also the question of who is 'any person' to be served with notice under section 82. Presumably, the principal target is the landlord's managing agent, who at least can seek to recover his costs for complying with the notice from the service charge (the Act only specifies payment of a fee for copy documents). The much broader description of 'any person' however may lead an RTM company to seek information from a range of other sources: possibly insurers or surveyors supervising particular works; even perhaps porters, caretakers or security staff who might carry contact details for leaseholders. Most likely recipients of a notice would include the landlord's solicitors or accountants, who will be most reluctant to hand out addresses and who would certainly look to charge a fee, probably to the landlord.

There is great scope here for confusion. It may well be that the courts will be called upon to interpret 'any person' more definitively in due course. Meanwhile, RTM companies will be likely to avoid the potential costs and delays caused by this provision where they can obtain the necessary information by different means, perhaps preferring to wait until they can use the rights in section 93.

That may be a luxury which is not available however to an RTM company which has yet to achieve the requisite 50% of qualifying tenants. Especially in developments with high proportions of sub-let flats, the landlord or his agent may be the only source for information on the whereabouts of leaseholders, leaving little choice than to serve notice under section 82 upon the landlord or agent, and taking the risk of alerting the landlord to the move to Right to Manage.

The alternative right to information under section 93 is wider, but of no help in the early stages. Notice may be given under section 93 to a landlord (of all or part of the premises), a third party to a lease (such as a management company) or a manager appointed by the court or the LVT under Part 2 of the Landlord and Tenant Act 1987. The notice, which is not in a prescribed form, requires the recipient to provide the RTM company with information in his possession or control which the company reasonably requires in connection with the exercise of RTM. The right to serve the notice arises 'where the right to manage premises is to be acquired by a RTM company' (section 93(1)), which logically must mean after service of the claim notice, or even after the time for a counter-notice has elapsed or the validity of the claim has been determined otherwise.

As with section 82, the notice may require the recipient to permit 'any person authorised to act on behalf of the company' to inspect documents and records in a readily intelligible form, and to supply copies (section 93(2)). Unlike section 82, there is no mention at all of payment of a fee (we shall look later at the costs of RTM).

The recipient must comply with the notice within 28 days of the date when the notice was given (it is not clear whether this means the date sent or the date received), but that is subject to the reservation that the recipient need not respond before the acquisition date (section 93(3)). To maximise its efficacy therefore, the notice should be served at least 28 days before the acquisition date. To avoid further delays, it would be wise to draw up the notice carefully, specifying as clearly as possible the information required.

The last thing the RTM company needs is to have to wait for lengthy periods after the acquisition date for essential information and documents, simply because the landlord or manager has room to quibble or seek clarification.

Right of access

In addition to the rights to documentation and records, section 83 grants an RTM company a right to obtain physical access to the subject premises. This right accrues following service of the claim notice, and extends to any part of the premises 'if that is reasonable in connection with any matter arising out of the claim to acquire the right to manage' (section 83(1)). What is reasonable may have to be determined by the court on an enforcement application under section 107, but as a rule of thumb it will probably be held reasonable to inspect the common parts but unreasonable (without special reasons) to gain access to demised flats or other units. The latter would risk putting the landlord in breach of his quiet enjoyment covenants, and may lead to actions for trespass.

The right of access becomes available not just to the RTM company. The persons who may utilise the right are:

- Any person authorised to act on behalf of the RTM company.
- Any person who is a landlord under a lease of all or part of the premises, or their authorised agent.
- Any third party to a lease or their authorised agent.
- Any manager appointed under Part 2 of the Landlord and Tenant Act 1987 or their authorised agent.

The right may be exercised on at least 10 days' notice to the occupier of the relevant premises or (if unoccupied) to the person entitled to occupation.

Perhaps the greatest benefit of the right of access for an RTM company would be to evaluate the state of repair and maintenance to the structural and common parts of the premises, especially if its authorised agent is a surveyor. Thus, the company could have notice before the acquisition date of the likely need for substantial expenditure once RTM has been acquired formally. This would assist budgeting, and may even have the effect of discouraging the exercise of RTM. The RTM company must bear in mind that it becomes responsible for repairs and maintenance immediately from the acquisition date, and will not want to be in the position of having no means to avoid early attack by dilapidations claims.

Until the acquisition date, the landlord remains obliged to repair, and what is revealed on an inspection may avail the RTM company's members (as leaseholders) to make a pre-emptive strike with a disrepair claim.

This is yet another of the strategic decisions to be considered by the RTM company between the claim notice and the acquisition date, and represents another telling reason for taking professional advice and assistance.

Checklist

The steps which need to be considered for the purposes of serving the claim notice and obtaining the necessary information for exercising the RTM may be summarised as follows:

1. Has sufficient time elapsed since service of the Notice Inviting Participation?
2. Have at least 50% of the qualifying tenants become members of the RTM company?
3. Consider serving notice under section 82 to obtain the necessary information for the claim notice.
4. Calculate the optimum acquisition date.
5. Prepare the claim notice (using the prescribed form) and ensure it is served upon the correct parties at the correct addresses.
6. Copy the claim notice to all qualifying tenants.
7. Consider serving notice requiring information under section 93.
8. Consider making use of the right of access under section 83.

Chapter 3

The Landlord's Response

In the first two chapters, we have looked at Right to Manage principally from the point of view of how leaseholders may exercise the right. This is inevitable, because all the initial steps are up to the leaseholders. The landlord (and his agents, contractors and so forth) may not know that RTM is proposed until the claim notice has been served, although he may have had an inkling earlier, especially if information has been requested under section 82.

Chapter 4 will set out what the landlord may do by way of a formal response, but in this chapter we shall consider the tactical options open to the landlord, the reaction of his agents and other contractors, and what steps must be taken in relation to existing contracts.

The landlord's response

The first decision for the landlord is whether or not he intends to oppose or admit the claim to RTM. As we shall see in Chapter 4, his choices are extremely limited, but there are steps which can be taken to oppose RTM, if not to prevent it. The landlord of course may be a residents' management company ('RMC'). There is no bar to RTM being exercised if there is already an RMC in existence.

Whatever the nature of the landlord, similar factors have to be examined. For example, if RTM is proposed in relation to one building on an estate of two or more blocks, the effect of RTM will be extremely disruptive and lead to considerable complexity in the continued management of the remainder of the development. A landlord which is made up of individual leaseholders or generally has the interests of leaseholders at heart must be concerned by the possible consequences upon the flat-owners as a whole by the exercise of RTM by as few as half of the leaseholders of one block.

At the other end of the scale, the more cynical landlord may be happy for a leaseholder company to take over the burden of his covenants for a while, particularly if he knows that major repairs are required. He could plan to sit back and let RTM take its course, and then take proceedings for disrepair or otherwise act in such an obstructive way that the RTM company is effectively strangled. Such landlords will have to weigh in the balance the risks of losing control of their blocks, perhaps permanently in some cases, against the chance to bring some leaseholder companies to their knees and rule out the prospects of further efforts at RTM for a considerable time.

Most investor landlords will no doubt accept the futility of outright opposition, and will not regard a 'scorched earth' policy as a desirable or time-efficient approach. Their concern will be the protection of their investment in the long term and the mitigation of shorter term losses. From both points of view, they would be well-advised to act in a co-operative fashion towards the RTM company. They will see insurance as one of the key issues. If the landlord (or the managing agent depending upon the parties and property concerned) succeeds in persuading the RTM company to continue with the landlord's block buildings insurance policy, this is likely to benefit both sides: the landlord can retain all or part of his insurance commission (often the most lucrative short-term return on his investment); meanwhile the RTM company can secure cover at premiums and terms not often available on a one-off basis.

If, however, the landlord finds that he is unable to achieve a good working relationship with the RTM company, or suspects that that will prove to be the case, he still has a couple of tactical approaches he can employ. He could attempt to dissuade enough leaseholders from participating to reduce the numbers to under 50%; or he could simply 'cut and run' by offering to sell the freehold to the leaseholders and remove himself from the equation.

As far as RMC's are concerned, they have little choice than to try to head off RTM by the persuasive route. If the management of their block passes from them, their whole *raison d'etre* will disappear, unless and until RTM fails and they are left to pick up the pieces. Even in an estate with more than one block, the management problems will be considerable. They would be magnified of course if different RTM companies were set up for each of the different blocks.

Incidentally, it is by no means inconceivable that such a scenario could occur. Many RMC's are run by a small minority of lease-holders, simply because they are the only members who have the time or inclination to join the board. The occasional flare-up of civil war which results from, say, the need for major works could be all that is needed for enough leaseholders in one block to be motivated to try RTM.

The likelihood is that RTM will occur most frequently out of disputes, notwithstanding that the right is exercisable without fault on the part of the current manager. As a brief perusal of cases dealt with by the LVT will show, disputes often flow from misunderstandings, distrust and poor communication. Landlords (including RMCs) will stand a far better chance of surviving RTM by improving their communication skills.

The managing agent's reaction

While the landlord will retain an interest in the property after the Right to Manage has been acquired, the managing agent's role will cease entirely on the acquisition date unless he is kept on by the RTM company. In many cases, the RTM company members will perceive the managing agent as the embodiment of the reasons why the move to RTM started in the first place. A charm offensive therefore will be that much more difficult for the agent, but the incentive could be that much greater.

Many managing agents have taken a pragmatic view about Right to Manage: anticipating that they will lose some appointments and gain others. Obviously however, agents cannot afford to be blasé, and will need to work hard to stay involved in the developments which contribute most to their profit margins. In doing this they have a difficult task. In all the consultation papers concerning the service charge aspects of the Act, the government has acknowledged that the extra responsibilities imposed by the Act (on the accounting side for example) will lead inevitably to increased management costs and therefore higher agents' fees. (The consultation papers can be seen on the website of the Office of the Deputy Prime Minister (www.odpm.gov.uk). In a competitive market meanwhile, RTM companies will be looking for savings, so agents hoping to increase and retain business will be forced to seek out ways to offer greater value for money.

Generally, agents must look at such things as their systems, delegation to more junior staff, and out-sourcing. In relation to RMCs and RTM companies, they may have to try to negotiate arrangements whereby resident directors take a more active role in some management and administrative functions.

In each individual case where RTM is proposed, the managing agent will need to consider carefully and coolly whether the development concerned is worth fighting for. Indeed, that thought process should perhaps start earlier than that. Once a move to RTM is countenanced and the RTM company directors have their own ideas about who is to manage the building, it will be very difficult for the incumbent agent to restore his position. If, however, an agent has already examined his portfolio and decided which properties he wants to keep, he would be well-advised to act in a pro-active fashion to secure the loyalty of the leaseholders. This may involve distancing himself from his current landlord client, but that is the sort of choice which now faces managing agents.

Of course, the managing agent may take the view in an individual case that he has no wish to work with the burgeoning RTM company; he may even be relieved at the prospect of losing management.

In such a case, his thoughts will concentrate on two things: facilitating a smooth handover; and protecting his own position. As to the first, the requirements on the landlord (and therefore his agents) in the context of the transfer of management will become clearer later in this chapter and in Chapter 6. In any event, managing agents will generally be expected to comply with the provisions of the various codes of practice (RICS, ARMA and ARHM) which deal with transfers of management.

From the point of view of self-protection, the managing agent has a number of factors to take into account. He will not wish to become embroiled in any form of litigation after the acquisition date, so he will need to take steps to tidy up the service charge accounts and the property records generally.

Inevitably, there will be things left undone as management passes; there will be neither time nor probably the funds to initiate outstanding repair works. Consequently, the agent will be well advised to ensure that outstanding items are properly documented, with file records of reports to his principal and the leaseholders (when appropriate) by which the agent has communicated items which need seeing to and suggested scheduling and budgeting to allow for them.

Most importantly, the agent may well have been in a position as the named contracting party with suppliers of services to the building, some of which could be ongoing. Although these contracts may survive the acquisition date, the agent must assume they will not and ensure that early steps are taken to bring them to an end on or by the acquisition date, with any payments due being charged to the service charge fund. If they have not already done so, the agents should check their records to ensure that all contractors are aware that they were acting as agents for their principal who is responsible for any claims in the final analysis.

Apart from the very last point, many of the comments above concerning managing agents will also apply to landlords.

The position of contractors

The managing agent is generally in a unique position because he contracts directly with the landlord but communicates directly with the leaseholders, and meanwhile he will arrange contracts with other providers. The situation facing the landlord's other contractors is thus somewhat less complicated, but none the less they will also have to consider their stance when they are notified of a proposal for Right to Manage.

The variety of contractors involved with a block of flats can be enormous. They may include:

- Insurers and insurance brokers.
- Employed wardens, caretakers, porters, etc.
- Lift maintenance engineers.
- Entry-phone or other security providers.
- Builders, roofers, plumbers, decorators, etc. who may be engaged on one-off repair contracts or ongoing maintenance.
- Surveyors and structural engineers.
- Accountants.
- Gardening contractors.
- Cleaners.
- Even utilities suppliers.

The majority of these contractors may not be greatly affected by the exercise of Right to Manage in an individual building from the point of view of their overall business, but they will all have decisions to make when RTM appears over the horizon.

Will they wish to continue to provide services to an RTM company if invited to do so? Many contractors provide services to blocks of flats because of their much broader relationships with landlords or managing agents; as a result, they are able to offer economies of scale. Unless the connection continues, most probably through the retention of the managing agent, they are likely to be unwilling to enter into contractual relationships with small RTM companies who may not be able to meet their bills at any point in time. Even if they are willing, they will almost certainly be unable to offer the same deals, especially without payment up front.

What will be the effect on their business? This is a particularly relevant consideration for small businesses – typically gardeners and cleaners. They may be sole traders or very small companies or partnerships who depend on a few contracts from the same landlord or agent in their locality: indeed, the property concerned might conceivably be their core business. There is thought to be a large number of such firms who will find it difficult to replace the work if a contract is lost through RTM. They may be forced to decide whether to keep operating at all. Alternatively, if their business is based upon a handful of contracts awarded by the same landlord or agent, they may have to determine whether offering their services to the RTM company will prejudice their working relationship with their usual employer. At the same time, they may dread the prospect of working for a leaseholder company whose priorities and way of conducting the relationship could be very different from that to which they are accustomed.

What is the position of employees? It is not the purpose of this book to attempt to set out employment law. Those who employ staff directly at leasehold developments and the employees themselves are encouraged to seek specialist advice on their relative positions in the event of the exercise of the Right to Manage. The position will differ considerably if the employment contract relates specifically to a particular development (the purpose of which could be frustrated by RTM), especially if the terms of employment require residence. The picture will be complicated further if the employment covers an estate development and RTM is acquired in relation to part only of the estate. It may be necessary in such circumstances for a scheme to be set up whereby the employer provides the services of the employee to the RTM company on a sub-contract basis.

Are their fees secure? When they receive notification that RTM is pending, contractors should be able to calculate the value of the work remaining to be carried out before the acquisition date. They will doubtless wish to ensure that they receive payment for their work-in-progress, as well as what has already been completed, while the current management is still in place. In any event, managers should be encouraging them to get their bills in so that a full account can be taken at the acquisition date. For those carrying out routine and ongoing tasks, this should be a relatively straightforward process.

There will be a slightly more complicated position for the likes of a large, one-off contract for major works. Here, it will be more difficult to assess the precise point which will have been reached by the acquisition date. Contractors will not wish to run the risk of committing themselves to funding their own staff and their sub-contractors to a level which cannot be supported by the service charge income or by the finance available to the RTM company when it acquires management. Consequently, there will be a temptation for the contractors to wind down their efforts pending payment. This is particularly so where the anticipated cost of the works was the decisive factor in prompting the leaseholders to elect to exercise RTM (as is likely often to be the case).

The considerations of the contractors will inevitably be determined by commercial factors. The RTM company ideally should have taken a view before serving the claim notice as to which services and which providers they wish to retain. Having reached conclusions on that, the RTM company should have a strategy in mind for persuading those contractors to stay. Part of that process must involve knowing who the contractors are and how to contact them, which is where sections 91 and 92 of the Act come in.

Contractor notices

Section 91 provides that that where RTM is to be acquired and there is at least one 'management contract' in existence at the premises affected, both contractor notices and contract notices must be served under section 92. In order to ascertain to which contracts and which parties section 92 notices apply, section 91 goes to great lengths to clarify and define the relevant terminology.

A 'management contract' is not simply the landlord's agreement with his managing agent, but is any contract entered into by the 'manager party' (the landlord, a third party RMC or a manager appointed under Part 2 of the Landlord and Tenant Act 1987) with the 'contractor party' 'under which the contractor party agrees to provide services, or do any other thing, in connection with any matter relating to a function which will be a function of the RTM company once it acquires the right to manage' (section 91(2)). In other words, it covers contracts with managing agents and any of the other contractors listed above.

An 'existing management contract' is any such contract which was in force immediately before the 'determination date' or was entered into between the 'determination date' and the acquisition date (see Chapter 2). The 'determination date' will be examined more fully in Chapter 4, but essentially it is the date by which any questions surrounding the validity of the claim to RTM have been removed, and thus the acquisition date can be firmly fixed.

Specifically, the Contractor Notice must be given by the existing manager to each of his contractors. If the contract was in being before the determination date, the notice must be given on that date or as soon afterwards as reasonably practicable. If the contract was entered into after the determination date it must be given as the contract is entered into or as soon after as reasonably practicable (section 92(2)).

There is no prescribed form for the Contractor Notice, but it must contain the following:

- Sufficient details to identify the contract.
- A statement that an RTM company is to acquire the right to manage the premises.
- The name and registered office address of the RTM company.
- The acquisition date.
- Advice that the contractor, if it wishes to provide the same services to the RTM company, should contact the RTM company at the address given.

There are two points which need to be made about the Contractor Notice.

First, the only address for the RTM company which will be notified to the contractor is that of its registered office. As with many small companies, this may be its accountant/solicitor or company agent, and thus may not facilitate quick or reliable communication. To a certain extent this problem can be overcome by the Contract Notice (see below), but it is a point worth bearing in mind when the RTM company selects its registered office.

Secondly, there is no requirement for the manager party to notify the contractor that the contract will terminate on the acquisition date. The government's view throughout the consultation process on the Bill and the passage of the Act has been that existing management contracts will be frustrated by the acquisition of RTM. (The consultation papers can be seen on the website of the Office of the Deputy Prime Minister (www.odpm.gov.uk). Whether or not that is correct (and it is not intended to argue the point here) it would be unwise to leave unsaid the consequences of the Contractor Notice. Either with the notice or before it, the manager party (whoever that may be) would be well advised to issue an unequivocal notice of termination to its contractors, expiring on the acquisition date. For accounting purposes, it may be useful also to ask the contractors to notify the manager of any outstanding costs and work in progress likely to accrue by the acquisition date (see

Chapter 6 as regards the manager's duties under section 94).

The recipient of a Contractor Notice is given a duty by section 92(4) to send a copy of the notice to any sub-contractors (and they, in their turn, must send copies to their own sub-contractors, and so on *ad infinitum*). Furthermore, they must send a Contract Notice in relation to any sub-contracts to the RTM company. (Section 92(5) defines 'existing management sub-contracts' in broadly identical terms to the definition of a 'management contract' in section 91(3). Likewise, the circumstances requiring notification of a sub-contract set out in section 92(6) mirror those for management contracts in section 92(2).)

It can be anticipated that some contractors will neglect to comply with these requirements, given that their level of interest will be lower than that of the parties to the leases, and especially as the Act is the only formal notification to contractors of these duties; they are not required to be set out in the Contractor Notice. Although they will be subject to possible enforcement measures through the county court under section 107, it would be more practical if the manager parties built into the Contractor Notices a warning of these provisions. This is a useful service which could be supplied by professional managing agents for example. Indeed, such a pro-active step may bring credit to the manager with the RTM company and contractors alike.

Contract Notices

A Contract Notice must be given by the manager party to the RTM company (section 92(1)), and this is also to be served on the determination date (or, if regarding a contract entered into after the determination date, on the date it was entered into) or as soon as reasonably practicable thereafter.

There is no prescribed form, but under section 92(7) the Contract Notice must include the following:

- Particulars of each management contract or sub-contract and of each contractor party.
- The address(es) of the contractor or sub-contractor concerned.
- Advice that the RTM company, if it wishes to take up the services of the contractors or sub-contractors, should contact them at the address(es) given.

This provision pre-supposes that the manager party has the requisite details of all the sub-contractors, which is perhaps optimistic. However, so long as the contractors comply with their duties to give Contract Notices, there will be little to be gained by seeking to enforce this against the manager. Once again, the manager is given an incentive to warn the contractors to provide this information.

Checklist for landlords/managers

1. Consider strategic response to claim notice.
2. Identify contractors.
3. Identify sub-contractors (where possible).
4. Consider giving notice terminating existing contracts.
5. Take advice on the position regarding any employees.
6. Warn contractors of the need to give Contract Notices concerning sub-contracts.
7. Prepare Contractors Notices, incorporating clauses to cover 3 and 4.
8. Prepare Contract Notice for service upon the RTM company.

Checklist for RTM companies

1. Consider at an early stage which, if any contractors' services should be retained.
2. Have alternative contractors in place ready for the acquisition date, especially for essential services.
3. Check that the buildings insurance will remain on foot after the acquisition date. If renewal is due around the acquisition date, ensure cover is readily available.
4. Ensure an appropriate address for the RTM company is given by the landlord/manager to contractors.

Chapter 4

Counternotices and Withdrawal

In the previous chapter, we have looked at some of the strategic and tactical considerations for landlords and managers when notified of a claim to exercise the Right to Manage. The opportunity for a formal response comes in the form of a counternotice under section 84 of the Act.

Counternotices

Counternotices generally will give the party who has received an originating notice the opportunity to express opposition, with reasons. This is *not* the case with a counternotice to a notice of claim for RTM. Right to Manage is operable without proof of fault; therefore, there are no grounds to oppose it on the merits of the claim – only its validity.

Even so, the prospects for opposition on validity are limited because a claim notice is not invalidated by any inaccuracies in its particulars (section 81(1)). Consequently, the recipient of a section 79 claim notice who wishes to place obstacles in the path of RTM may have to be imaginative in the drafting of a counternotice.

Service of the section 84 counternotice is not mandatory, but if it is served it must be done by the date specified for the purpose in the claim notice (not earlier than one month after the claim notice – section 80(6)). The manager party therefore may have as little as a month to consider its response. If this is an RMC, this will require the convening of a board meeting with relatively little notice. When it meets, the board will have to determine what response best serves the interests of the company members (who will presumably include the RTM company members). This could create something of a dilemma for the board, who will have to account to the full membership of the company in due course.

The counternotice can do only one of two things. Either:

- Admit that on the 'relevant date' (the date of the claim notice) the RTM company was entitled to exercise RTM in respect of the specified premises; or
- Allege that the RTM company was not so entitled on specific grounds under the RTM provisions in the Act (section 84(2)).

It is important to note that the question of entitlement revolves around the date the claim notice was given. The claim would not be invalidated if, for example, one or more RTM company members sold their flat shortly after service of the section 79 notice, even if the remaining members' interests then added up to less than the required 50% of qualifying tenants. The manager party could not raise in opposition that the RTM company represented a minority of leaseholders by the time of the counternotice. Neither does RTM become invalid later even if the company membership were only a small minority by the time of the acquisition date. This would be a matter for the RTM company to decide whether they should carry on in the circumstances.

The counternotice is in a prescribed form, set out as Schedule 3 to The Right to Manage (Prescribed Particulars and Forms) Regulations 2003. If the recipient of a claim notice intends to allege lack of validity on statutory grounds, he must specify which provisions of the Act have not been complied with. Perhaps surprisingly, there is no place on the form for a statement supporting the allegation. For example, the recipient could allege that particulars of the leases are cited incorrectly without saying why. Even though the Act provides that the claim notice will not be invalidated by any inaccuracy in its particulars, a counternotice alleging invalidity will still be able to cause delay at least.

Of course, one would hope that landlords will not manufacture doubts over validity of a claim to RTM simply to be obstructive, but such an approach is conceivable, and the best that RTM companies can do to prevent it is to make sure they have taken all possible steps to prepare their claim notices properly.

The other points to be included in the counternotice under section 84 are:

- A statement that, if the counternotice has alleged invalidity, the RTM company may apply to the Leasehold Valuation Tribunal for a determination whether the company was entitled to exercise RTM on the date of the claim notice (section 84(3)).
- A statement that, in the same circumstances, the RTM does not accrue unless the LVT has determined the issue of entitlement or the person giving the counternotice has agreed in writing that the RTM company was entitled to exercise the Right to Manage the premises (section 84(5)).
- The notes to the prescribed form, which include references to section 84(7) and (8) and, most importantly, the information

that an application to the LVT must be made within two months of the counternotice (section 84(4)).

If more than one counternotices have been served (as could be the case for example if there were a landlord and an RMC as a third party to the leases), the time-limit for an application to the LVT runs from the date of the last counternotice to be given.

The effect of a counternotice

Either a counternotice which admits the entitlement to RTM or the absence of a counternotice altogether will mean that RTM can proceed unchecked and the acquisition date will be as stated in the claim notice.

Since there appears to be no means by which the recipient of a claim notice can seek an extension of the time within which to serve the counternotice, it would seem that late service will not count and the effect will be as though no counternotice had been served.

To summarise the position if a counternotice is served which alleges that the claim notice does not comply with the provisions of the Act:

- The RTM company may apply to the LVT for a determination that it is entitled to exercise RTM (section 84(3)).
- Any such application must be made within two months of the last counternotice given (section 84(4)).
- If no application is made within two months, the claim notice is deemed to have been withdrawn (section 87(1) – see below).
- The company does not acquire the Right to Manage unless and until the claim is determined by the LVT or the person(s) giving the counternotice(s) has agreed the claim in writing (section 84(5)).
- If an application to the LVT is dismissed, the claim notice ceases to be of effect (section 84(6)).
- The LVT's determination becomes final when any appeal is disposed of or, if there is no appeal, when the time for appealing has expired (section 84(7)).
- An appeal is disposed of if it is determined and the time for further appeal has passed, or if the appeal is abandoned or 'otherwise ceases to have effect' (section 87(8)).
- In any of the above cases there will be an effect on the acquisition date (see Chapter 5).

Landlords who cannot be traced

There will be occasions when the landlord of premises cannot be found or identified. Section 85 provides that the RTM company may apply to the Leasehold Valuation Tribunal for an order permitting the company to acquire the Right to Manage in such circumstances.

Clearly, no claim notice can be served on an absent landlord, but there are other pre-requisites to an application:

- The qualification criteria for RTM must have been met.
- Notice of the application must be given to all qualifying tenants of flats in the premises (which might conceivably include an associated company of the landlord).

In dealing with the application, the LVT may issue directions, such as:

- The RTM company may be required to take further steps (advertisement for example) to trace the landlords or third parties to the leases (section 85(4)).
- If any parties are traced, no order will be made on the application, but the LVT will determine the rights and obligations of all affected parties as though a claim notice had been served and give such directions for implementation as it sees fit. In these circumstances, the LVT has power to modify or dispense with any of the statutory provisions in relation to RTM (section 85(5) and (6)).
- If the previous paragraph applies, the original application may only be withdrawn with the consent of the parties traced or the permission of the LVT (section 85(8)); and permission will only be given if it appears just to the LVT to do so in the light of information gained by the RTM company through tracing the parties concerned (section 85(9)).

In cases such as these, the service of a claim notice is not required, and so there is no room for a counternotice. No doubt the manager party concerned will be able to make representations to the LVT as to the validity of the claim to RTM as part of the process.

Withdrawal of the claim

Withdrawal of a claim to RTM may be voluntary or deemed to have occurred.

If voluntary and after service of the claim notice, the RTM company must give a notice of withdrawal under section 86 to:

- Each landlord under a lease of all or part of the premises.
- Each third party to a lease.
- Any manager appointed for the premises under Part 2 of the Landlord and Tenant Act 1987.
- The qualifying tenant of any flat in the premises.

The notice can be given at any time before the acquisition date, so possibly after Contractors Notices have been served. It must be deduced that the existing manager, who will remain in place, has the task of informing contractors of the withdrawal. The question of costs is dealt with below, but the Act makes no mention of any liability on the part of the RTM company or its members for compensation or increased contractual charges which result from the disruption to existing contracts. The manager will no doubt have to pass such losses on to the service charge payers, whether or not they participated in the claim to RTM.

In due course, one or more service charge payers may challenge the reasonableness of these charges through the LVT. The LVT will have no jurisdiction to apportion blame as between the manager and the RTM company, so it would seem likely that any reduction ordered by the LVT will have to be met by the manager. This would seem a particularly unfortunate outcome for the likes of an RMC.

Deemed withdrawal

Section 87 sets out the circumstances in which the claim notice will be deemed to have been withdrawn:

- Where a counternotice is served under section 84 questioning the validity of the claim and no application is made by the RTM company to the LVT to have the claim determined, or an application is made but subsequently withdrawn. If no application has been made, withdrawal will be deemed to have occurred at the end of the period allowed for the application (two months from the last counternotice); otherwise it will be the date of withdrawal of the application (section 87(2)).
- The previous paragraph will not apply however if the person who gave the counternotice subsequently agreed the entitlement to RTM in writing (section 87(3)).

- In the event of a winding-up order or an administration order affecting the RTM company, or if the company passes a resolution for voluntary winding-up.
- A receiver or manager is appointed in respect of the company's undertaking, or possession is taken by debenture holders of any property charged by the RTM company.
- A voluntary arrangement is approved in respect of the RTM company under the Insolvency Act 1986.
- The RTM company is struck off the register under section 652 or 652A of the Companies Act 1985.

A number of these events, and others, would also lead to the cessation of RTM after the acquisition date, as we shall see in Chapter 8.

Costs

It will not have escaped the attention of property managers that landlords and managers of various types will have a good deal of work to do on receipt of a claim notice. They need to consider the validity of the claim and their response to it; they are required to prepare and serve Contractor Notices and Contract Notices; they will have information to supply and accounts to make up; and quite apart from the procedural steps laid down in the Act, there will inevitably be a substantial amount of time spent in liaising and communicating with all concerned, including contractors and the non-participating qualifying tenants who will be looking to their own position in the light of RTM.

Consequently, costs will be incurred, possibly in considerable amounts. Although it is unlikely that managers will be able to recover their entire outlay, the Act does provide some degree of comfort. From the point of view of the RTM company and its members, they need to be aware of the extent of their liabilities in costs when weighing up the large number of factors which flow from the exercise of the Right to Manage.

Section 88 provides that costs are payable in the following circumstances:

- An RTM company is liable to any landlords of all or part of the premises, a third party to a lease, and any manager appointed under Part 2 of the Landlord and Tenant Act 1987 for their 'reasonable' costs incurred following service of a claim notice.
- Professional fees are to be regarded as reasonable only to the

extent that the receiving party could have been reasonably expected to have to pay such costs personally (section 88(2)).
- An RTM company will also be liable in respect of costs incurred by the manager party before the LVT when the entitlement to RTM is determined, but only if the application is dismissed (section 88(3)).
- If costs cannot be agreed, they are to be determined by the LVT.

Section 89 deals with costs in the event that RTM ceases, including by withdrawal or deemed withdrawal, as follows:

- The liability under section 88 continues up to cessation of RTM.
- Each member of the RTM company (past and present) is jointly and severally liable with the company for such costs (section 88(3)), unless he has since assigned his lease and the assignee has joined the company (section 88(4)).
- An operative assignment includes an assent by personal representatives and an assignment by operation of law to a trustee in bankruptcy or on foreclosure of a leasehold mortgage under section 89(2) of the Law of Property Act 1925.

Given that an RTM company is effectively a 'man of straw', the provision that each member is jointly and severally liable for the manager's costs if RTM ceases will be reassuring to the manager parties and, conversely, a risk which should be advised to leaseholders when contemplating joining the company. Indeed, such a warning is required in the Notice Inviting Participation under section 78.

No mention is made of the costs of a counternotice. Theoretically, there is no bar to the manager claiming his costs of a counternotice even if it was subsequently withdrawn (and even if it was used simply as a delaying tactic) on the basis that these were costs incurred 'in consequence of a claim notice' (to use the form of words in section 88(1)). It is highly probable however that such costs would not be held as reasonable if they fell for determination at the LVT.

There may also be scope for argument at the LVT if the manager seeks to put any shortfall in his costs to the service charge. This would have to be the subject of a separate application under the LVT's service charge jurisdiction, although no doubt it could be heard at the same time as any other issues more directly concerned with the Right to Manage.

Where the manager party is an RMC, they will have little choice than to look for maximum costs recovery from the RTM company or the service charge, because they have no alternative source of funds.

In any event, it can be expected that there will be litigation on costs questions generally, and it is to be hoped that case law assists in drawing lines in the sand to distinguish between what costs may or may not be recoverable in due course. Until that happens, RTM company members should be aware not only of the risk of personal liability for costs, but also of the prospects that costs claimed could be substantial, and that arguments over costs may lead to further expensive satellite litigation.

Chapter 5

Acquisition

Acquiring the right

Although qualification for RTM is relatively straightforward, and achieving the right can be done without having to prove fault on the part of the current manager, it can be seen that following the procedural steps requires a great deal of hard and pains-taking work by the RTM company – especially if the landlord chooses to be obstructive.

Reaching the acquisition date (when the RTM crystallises) may take many months, and RTM company directors (and those advising them) should be alert to the possibility that some members may have been tempted to fall by the wayside. Indeed, some members may have sold their flats to new leaseholders who have little or no knowledge of the move to RTM or the reasons behind it. The RTM company will have been wise to ensure constant channels of communication with company members and other qualifying tenants who have yet to join.

It would also be prudent to keep the original motivation for RTM at the forefront of everyone's minds, including the company's directors.

The acquisition date

In Chapter 2 we looked at the calculation of the acquisition date when preparing the section 79 claim notice. Then in Chapter 4 we saw how the originally planned date can be affected by a landlord's counternotice. It is now worth summarising the various ways in which the acquisition date can work out in practice. These are set out in section 90 of the Act:

- If no counternotice has been served disputing the entitlement to RTM, it will be the date originally spelt out in the claim notice (ie at least four months after the claim notice).
- If a counternotice disputing entitlement has been served, but the person giving it subsequently agreed in writing that the RTM company was entitled to proceed, it will be three months after that written agreement.

- If the question of entitlement had to be determined by the LVT, it will be three months after that final determination.
- Where the LVT had to make an order in the event of an untraceable landlord under section 85, it will be such date as is specified in the order.

Thus, the minimum possible period from service of the claim notice to the acquisition date is four months (the claim notice itself is unlikely to be served for a month at the very least from the inception of the RTM company). The maximum possible period is impossible to fix as it will be dependent upon the speed of the LVT, and that will be affected to a certain extent by the efficiency of the RTM company and the obduracy of the current manager. The speed of the LVT varies somewhat according to region, but the track record suggests that a case could take anywhere between six months and 18 months, or even two years. It would not be surprising then to find that a disputed claim to RTM could work out along the lines of the following example:

- 1 May – claim notice served, giving the recipients until 1 June to serve a counternotice and fixing 29 September (the nearest quarter day) as the putative acquisition date.
- 31 May – the landlord serves a counternotice disputing the entitlement to RTM. The RTM company has two months within which to apply to the LVT to determine the issue (section 84(4)).
- 14 June – the company issues the LVT application, having taken 14 days to put the papers together and gather the application fee.
- 14 December – the LVT's final hearing.
- 14 March – the acquisition date (three months after the determination by the LVT).

It has to be said that the above timetable is somewhat optimistic (from the point of view of the RTM company), but even so, it has taken over 10 months from the claim notice and probably a full year from the decision to launch the RTM company. During that time of course, the landlord will have continued as manager and carried through his plans for the year's service charge expenditure (subject to any applications to the LVT on reasonableness). If the initial motivation for many RTM company members was sparked off by the landlord's proposals for substantial expenditure, they will be wondering why they did not simply go to the LVT to determine the expenditure levels.

This scenario underlines the point that RTM is by no means an easy way of changing management and reducing expenditure in the short term. On the other hand however it will provide that opportunity in the longer term, and should save the need for regular annual LVT cases concerning the service charges raised by the current landlord.

The potential for delay which is built in to the procedure for acquiring the RTM (at least in the event of an obstructive landlord) is another factor to be taken into account by leaseholders when considering if RTM is indeed the strategy which will best suit their interests.

Among the other parties who could be affected by uncertainties concerning the acquisition date are:

- The landlord.
- Individual leaseholders, especially if they intend to sell their flats.
- The managing agents and other contractors.

The landlord

Although the landlord may have contributed to the delay and uncertainty surrounding the acquisition date, he could have done so from perfectly proper motives. Either way, the landlord will suffer problems. He cannot afford to assume that the exercise of RTM will prove to be successful in due course and, unless and until management is transferred to the RTM company, he will still be obliged by his covenants under the leases to continue to manage, maintain and repair the premises.

While carrying out his responsibilities, it is quite probable that the leaseholders will be more resistant than usual, whether or not they are participating in the RTM company. Cash flow may slow down or even dry up. This will present particular difficulties to RMC landlords or managers.

Meanwhile, the landlord will be reluctant to risk committing himself to substantial expenditure or new projects with the imminent threat of RTM hanging over his head like the sword of Damocles. He will not want to compound the problems he has already with existing contracts (see Chapter 3), nor will he wish to provoke new arguments with leaseholders.

For most landlords therefore, there is an advantage to knowing the acquisition date as early as possible. This may provide an incentive to admit the claim to RTM without demur; alternatively, landlords may prefer to be in control of the timing of the acquisition date which can be achieved by serving a counternotice but subsequently withdrawing it. The acquisition date will then be three months after the landlord's written withdrawal of his dispute (section 90(5)).

Individual leaseholders

While the effects of uncertainty upon the landlord raise significant problems of a largely practical nature, the consequences for leaseholders may be more far-reaching.

All leaseholders, whether or not they are participating in the RTM company, will want their investment protected. It is therefore in their interests for the current manager to continue to ensure that the premises are properly insured, repaired and maintained. At the same time however leaseholders will be tempted to hold back from contributing funds to a management regime which appears to be on the way out.

Leaseholders who are attempting to sell their flats will have special problems. Purchasers are likely to be put off if they cannot have satisfactory assurances about continuing management, or the likely future costs of service charge matters, or even who is going to be the party responsible for management in the future. Moreover, purchasers who need mortgage finance will find lenders reluctant to advance funds on a property whose maintenance, repair and insurance provision is in a state of flux.

Subject to market conditions to a certain extent, vendors could find themselves losing buyers or having to reduce the selling price to achieve a sale during the period of uncertainty. At the very least, purchasers are likely to be advised to seek more than usual in the way of retentions and indemnities.

Such factors may conceivably lead to some leaseholders wishing to withdraw from RTM in some cases, especially if a counternotice has been served under section 84. Where the majority for RTM is already slight, this could lead to the collapse of the RTM company and thus the withdrawal of RTM. (The more cynical landlords may even prey on nervous vendors.) It is not that simple however. Article 12 of the prescribed articles of association requires seven clear days' notice of resignation from the company, but further provides that notice will not be effective if it is given between the date of the claim notice and the acquisition date or the date of withdrawal or deemed withdrawal of the claim.

From a legal point of view therefore the RTM company's members are locked in, but clearly this does not preclude them from expressing a view, the force of which will be dependent on numbers.

The managing agents

Whatever the nature of their landlord client (corporate landlord, investor or RMC), managing agents will be concerned to continue providing a professional service. Assuming the agents abide by the relevant code of practice (RICS, ARMA, or ARHM), they will be conscious of their duties both to their clients and to the leaseholders. Most managing agents will also be aware of their and their clients' fiduciary duties as trustees of the service charge fund (section 42, Landlord and Tenant Act 1987). Both as trustees and under the various codes of practice (and indeed under the 2002 Act), agents are required to preserve the funds and to co-operate in a smooth handover.

Accordingly, managing agents will have the same concerns as their clients regarding continuing management. In addition, agents are contractors and will need to consider their own business: are they looking to be retained by the RTM company; do they run the risk of alienating their existing client; how should staff best be deployed to continue to service the premises while seeking replacement business; how do they maintain their profit margins? In planning their strategies, managing agents will also suffer the consequences of the uncertainties of timing and the eventual outcome.

Other contractors

We have already discussed the position of contractors in the light of impending RTM in Chapter 3. Their approach will vary widely depending on their size, the degree to which they regularly supply services to the premises concerned, and the importance of the contract to their business. The more concerned they are by the effects of RTM, the more they will face disruption on account of delays and uncertainty over the acquisition date.

If the contractors have already agreed terms to continue supplying services to the RTM company after the acquisition date, they will be less disturbed. Even so, they will need to know in due course precisely who they are working for and when – particularly if their charging structures are to change.

Contractors who have received notice that their contracts are to be terminated will need to know the exact date of termination. If the claim to RTM is to be referred to the LVT, it will be impossible to fix the acquisition date with any accuracy for some considerable time; however, the contractors may insist that the current manager gives them a firm date when the Contractor Notices are issued, or they may determine their own date. There is therefore a strong possibility that a number of contractors could have withdrawn their services for a substantial period before the acquisition date, leaving a hole in the services which the manager is able to provide.

Alternatively, if a date is set which transpires to be beyond the acquisition date, the current manager may find himself contractually bound to pay the contractor for a period when he has difficulty in recouping his expenditure from the service charge fund. In the case of an RMC landlord or manager, the inability to meet contractual obligations could lead to insolvency (with potential consequences for RMC directors personally if they are deemed to have allowed the RMC to trade while they knew or ought to have known that the RMC would be unable to cover its debts). In any event, the contractor may end up out of pocket in real terms. For smaller contractors, this could lead to their own insolvency.

Contractors with greater bargaining powers (and access to legal advice) might look to the current manager for indemnities, possibly even before RTM is contemplated. Once again, RMC landlords or managers could face particular problems.

In the light of all these factors, landlords, RMC's and their directors, agents and contractors may be looking to the insurance industry for new forms of cover for the variety of potential liabilities introduced by the Right to Manage.

Novation

Whether as a result of contact made after service of the Contract Notices under section 92, or following some other form of communication between the parties, there will be occasions when RTM companies come to agreements with contractors for the continuation of the contractors' services after the acquisition date. The process by which one contracting party's obligations to another are extinguished by the performance of those obligations to a substitute, with the express or implied consent of all concerned, is covered by the legal doctrine of novation.

It is not the purpose of this text to give a learned analysis of novation or its effects, but it may be as well for parties with substantial or multiple contractual arrangements which will be caught up in the exercise of the Right to Manage to obtain specialist legal advice on the consequences of novation upon their contracts. (The same could be said of the doctrine of frustration which is expected to apply to contracts which are terminated by the exercise of RTM.)

Turning instead to the practical side of these events, the RTM company may need to ensure that the contractual terms are clearly agreed and understood. To assist in this process, the company could seek information of the current manager's understanding of the existing terms by notice under section 93 (see Chapter 2). It may be, for example, that there are side issues to the existing contractual arrangements (discounts, referral fees, etc.) which are outside the RTM company's power or control. This is particularly likely with managing agents, but also is perfectly possible with insurers and a whole range of other contractors.

Bearing in mind that there might be a relatively short time between the Contract Notices and the acquisition date, and the need for the RTM company to provide essential services immediately from acquisition, the company may need to act quickly to have their contracts in place.

Chapter 6

Rights, Duties and Responsibilities

From the acquisition date, the right to manage the premises passes from the landlord or manager to the RTM company. Neither the landlord's rights nor his duties come to an end there however. The freehold does not change hands, and the parties to the leases remain the same (including any RMC's, although their role is likely to become marginalised).

The landlord's rights and duties

Although the landlord has conceded the right to manage, he receives instead a new right to have a say in the management of the premises. This right is manifested in two ways:

- By section 74(1)(b) of the Act, any landlords under leases for all or part of the premises are entitled to membership of the RTM company from the acquisition date. This would therefore include the freeholder and any intermediate landlords under long leases, but would not include any other form of manager such as a manager party under a lease. An RMC therefore would be entitled to join the RTM company only if it owned the freehold or was otherwise a landlord under a long lease. (Where a developer sets up a structure by which the freehold is to pass to an RMC as soon as possible after the last unit has been sold – a common arrangement nowadays – this section gives an extra incentive to both the developer and the RMC to complete the transaction quickly. The developer will not wish to become involved in an RTM company, whereas it is in the RMC's interests to do so.) The voting rights in the RTM company are complex and will be examined in Chapter 7.
- At the same time, section 97(1) provides that the management duties which now become the responsibility of the RTM company (which will include many of the landlord's covenants) are owed to any landlords of long leases affecting the premises in the same way as to leaseholders. This provision will be

analysed more closely below and in Chapter 7, but the thrust of it is that a landlord will be able to take the RTM company to the LVT or sue for breach of covenant in relation to what were previously the landlord's own responsibilities.

As the freehold has not changed hands, the landlord retains the right to collect and receive ground rent, and the right of forfeiture (section 96(6)). However, he will be able to exercise forfeiture only in respect of covenants due directly to him (such as payment of ground rent) or if requested to act by the RTM company on covenants now due to the company (such as payment of service charge) (see Chapter 7). Having said that, the landlord will still be able to enforce breaches and arrears which accrued before the acquisition date.

In any event, the Act places further restrictions on the use of forfeiture (sections 167–171), which will be examined in detail in Chapter 16.

Of course, the exercise of RTM may not affect all of the premises within the ambit of the current manager. The Right to Manage covers only those parts of the premises which comprise or provide services to the qualifying flats; commercial units and non-qualifying flats remain the responsibility of the landlord.

Furthermore, a particular RTM company might have exercised the right in relation to a single building on an estate development consisting of two or more buildings, including perhaps some freehold dwellings which share and contribute to common services and amenities.

Each of these scenarios brings its own complications for landlords and RTM companies alike:

- Commercial tenants. Many blocks of flats with commercial elements will survive RTM without too much difficulty if the residential and commercial areas are managed entirely separately. This is not always the case however. There may be a common service charge fund for example, which will create problems for the accounting, although this could be overcome (with co-operation on all sides) by a single overall account comprising two otherwise separate accounting systems for the residential and commercial areas (the advisability of using the same managing agents and accountants for both parts is self-evident). Shared or overlapping common parts and services will be more problematic however. Not only might there be confusion over who is responsible for managing such areas, but there could also be conflict between the different priorities of commercial and residential tenants. To use redecorations as one example, it is quite probable that commercial tenants will be willing to pay a share of more frequent and higher quality work than residential leaseholders. The landlord will have to find a way of resolving this dichotomy if he is to avoid breaches of his covenants to the commercial tenants, and his rights as regards the RTM company may have to come into play.

- A large number of flats within blocks will not be held by qualifying tenants for a number of reasons. They may simply be empty, or perhaps let out on assured or regulated tenancies. Occasionally, they might be retained for the landlord's own use, or to house an employee such as a resident warden, porter or caretaker. The landlord is responsible for the service charge contributions for these flats, whether or not the leases so provide (section 103 – see below). The landlord will also retain the management functions in respect of them. If they are entirely self-contained individual flats scattered about the building, there should be few additional complications. If, however, they have a floor of the building to themselves, or the structure of the property is such that they have access to common parts and services which are separate from the qualifying flats under the RTM company's management, there will again be overlap in management responsibilities. The upshot will be similar to that affecting commercial elements (which will be exacerbated if there are commercial units and retained flats in the same building).
- The qualifying criteria for premises set out in section 72 (see Chapter 1) do not preclude the possibility that the Right to Manage can be exercised in respect of one building in an estate consisting of two or more blocks of flats (or indeed other buildings) sharing common parts, services and amenities. There is no requirement for the leases for the estate to be varied, nor is any variation implied by operation of the Act. The landlord (quite likely an RMC in such circumstances) will still be bound by his covenants to the leaseholders of the other blocks, including with regard to common parts which serve the RTM block. There is also very likely to be a single service charge fund for the whole estate. Quite who is responsible for maintaining the common services and amenities (such as estate roads), and who is to contribute and in what proportions are questions which both the landlord and

the RTM company will need to resolve at a very early stage. It may well be that variations of leases will be necessary, but co-operation will certainly be the top priority. Again, it would seem desirable to employ a single managing agent (although the agents may wish to consider the potential for conflicts of interests). The situation will be even more complex if there are different RTM companies for each block, as is perfectly conceivable under section 72. (It has been suggested that, in such circumstances, the RTM companies operating on the same estate could merge. It is not yet clear whether this will be permitted in the light of the contents of the prescribed memorandum and articles of association.)

- It has become commonplace for new leasehold developments to include a number of independent freehold units on large estates. This occurs frequently in retirement developments. The freehold householders are none the less connected to the leasehold areas because of their use of common services and amenities to which they contribute through some form of rentcharge or estate service charge. As freeholders, the householders have no standing in the landlord and tenant legislation (although they may now be able to apply to the LVT in some circumstances to determine unreasonable estate charges – see Chapter 12). Neither are they of a class entitled to membership of an RTM company. Accordingly freehold householders may be prejudiced by the exercise of RTM in part or most of the estate. The overall freeholder of the estate will still be obliged by covenants to the householders, and will probably be forced into the position of representing their interests (albeit in a limited manner) when using his rights in respect of the RTM company.

Inevitably, in any of the sets of circumstances outlined above, there will be scope for disputes between any or all of the various interested parties affected by the exercise of the RTM. Not all of them will have recourse to the LVT for dispute resolution (for example, the LVT has scant jurisdiction with regard to freehold householders and none for commercial tenants). In complex but common developments such as these, the parties would be well advised to consider setting up their own dispute resolution mechanisms, preferably incorporated into their leases or other documents of title (so as to carry on to subsequent assignees and purchasers). Managing agents could perhaps offer a service to this effect; their intimate knowledge of the properties and the intricacies of management would seem to put them in the best position.

In each and every property in which the Right to Manage has been acquired (whether or not any of the above apply) the landlord's or other manager's principal duty to the RTM company is in relation to service charges.

Service charge funds: landlord's contributions

With regard to non-qualifying units (whether commercial or residential) the landlord is obliged to contribute their share of service charges by section 103. The obligation arises when the contributions payable under the leases for the qualifying flats add up to less than 100% of the total service charge expenditure. This section does not deal with a situation whereby the leases for qualifying flats cover 100% of the service charge, notwithstanding that there are retained units which make no contribution. Such a circumstance would still require rectification of the leases by variations (which can now be ordered by the LVT – see Chapter 13).

The shortfall is payable by the 'appropriate person' as defined by section 103(5), as follows:

- The landlord under a lease for the unit.
- If there is more than one lease for the unit, the immediate landlord under the most inferior lease (so if there were a head-lease, an under-lease, and a sub-under lease, it would be the landlord under the sub-under lease).
- If there is no lease, the freeholder (for example, this could apply to any rentcharges payable to the overall service charge fund by freehold householders on the development).

In this context, 'lease' (as opposed to 'long lease', defined by sections 76 and 77 and referred to in Chapter 1) includes a tenancy: for example, a business tenancy (section 112(2) contains the definitions).

This provision seems simple enough, until it tackles a situation with more than one 'excluded unit' (as it is termed by section 103). Then the appropriate person for each unit will pay a proportion of the shortfall which is ascertained by computing the ratio of the unit's internal floor area to the internal floor area of the excluded units as a whole (section 103(4)). This statutory formula will override anything in the leases or tenancies for the units concerned.

From the point of view of the RTM company, the way in which

the contributions to the shortfall are apportioned will be of no more than academic interest, so long as the service charge contributions are brought up to 100%. The landlord(s) and the tenants of the excluded units may look at it differently however. For example, they may consist of a mix of commercial and residential units. The commercial tenants would expect normally to make a higher contribution pro rata. Moreover, the statutory formula takes no account of the fact that some units may have within their demise quite substantial and valuable external floor areas.

In more complex developments, it may be perceived as unjustifiably onerous on some parties to abide by the formula in section 103(4), and the landlords and tenants of such units may be tempted to seek alternative ways around the apportionment question.

Service charge funds: accrued uncommitted service charges

Section 94 of the Act imposes a duty upon the current holder of service charge funds (which could be the landlord, a third party manager such as an RMC, or a manager appointed under Part 2 of the Landlord and Tenant Act 1987) to pay to the RTM company any 'accrued uncommitted service charges' standing to the account on the acquisition date.

The accrued uncommitted service charges are to be calculated in accordance with section 94(2) as follows:

1. Compute any sums paid by way of service charges in respect of the subject premises.
2. Add any investments which represent such sums, including income earned on them.
3. Deduct any amount which may be needed to discharge costs incurred before the acquisition date which are proper items of service charge expenditure.

In other words, what is payable is any surplus carried forward from previous years, plus contributions paid on account in the current year, plus any reserve fund, less what is required to pay current and known bills to the service charge. (The calculation will be less straightforward in respect of a service charge fund covering an estate where RTM only affects part of it.)

If the amount cannot be agreed or readily ascertained, either the outgoing manager or the RTM company may apply to the LVT to determine the amount (section 94 (3)). This would obviously include any dispute about what is or is not an item of expenditure properly attributable to the service charge.

Once the figures have been ascertained, the duty to pay over the fund must be complied with 'on the acquisition date or as soon after that date as is reasonably practicable' (section 94(4)). Of course, if the LVT has to determine the amount, it may be several months before the final figure is known. No doubt the LVT will encourage the outgoing manager to transfer as much as possible on account.

The duty under section 94 effectively requires the manager to take a full account of the service charge fund as at the acquisition date, for which purpose managers will need to ensure that they are aware of any potential liabilities to the fund up to that time. In particular, managers will want to avoid being caught out by an unexpected bill after they have handed over the balance of the funds. It would be prudent therefore to request details of outstanding costs and work in progress from all contractors, possibly through an additional clause in the Contractor Notices under section 92 (see Chapter 3).

As a result of this exercise, it may well transpire that there are no funds or very little which can be passed over to the RTM company. The RTM company would still be able to apply to the LVT for a determination under section 94(3), but that could prove to be fruitless, and anyway will cost time and money. Meanwhile, the RTM company would be bereft of service charge funds to carry out its new management functions in this scenario. Unless the company has acquired funding from other sources or its members are willing to lend it sufficient monies pending collection of subsequent service charges, this will put the RTM company in an extremely difficult position.

The question of alternative funding therefore is one which the RTM company should consider at a very early stage, and certainly well before the acquisition date.

Management responsibilities

On the acquisition date, the responsibility for performance of the landlord's management functions passes to the RTM company. The management functions which pass are those of the landlord of all or part of the premises (section 96(2)) and those of any third party manager such as an RMC (section 96(3)). Neither such party retains any management function while RTM continues to be exercisable in relation to the premises (section 96(4)).

What those functions consist of is set out in section 96(5) of the Act:

- Services
- Repairs
- Maintenance
- Improvements
- Insurance
- Management

In short, anything for which a service charge might be payable becomes the responsibility of the RTM company on the acquisition date, except matters which relate solely to flats or other units which are not qualifying flats (section 96(6)). Neither does the RTM company inherit any functions regarding the right of re-entry or forfeiture (section 96(6)(b)).

It appears to be possible for the RTM company to delegate management functions to the current manager. Section 97(2) provides that no landlord, third party under a lease, or manager appointed under Part 2 of the Landlord and Tenant Act 1987 may exercise any of the management powers passed to the RTM company, except with agreement between that party and the RTM company. For example, the RTM company may find it convenient to retain the landlord's existing insurance arrangements.

In any event, section 97(3) specifically allows any person to insure all or part of the premises. Without the RTM company's agreement however, the cost of such insurance could not be passed on to the service charge, and would thus be at the insured's expense.

Meanwhile, leaseholders' obligations to the landlord are owed to the RTM company from the acquisition date (in so far as they relate to the functions transferred to the RTM company) (section 97(4)); the most obvious example would be payment of service charge contributions. Section 97(5) however expressly excludes from this provision the payment of service charges in respect of costs incurred before the acquisition date. These payments still fall to be made to the landlord or other outgoing manager.

Approvals

One function which does not pass entirely to the RTM company is the landlord's right to grant approvals under the leases. Landlords retain unfettered the right to grant approvals in respect of flats or units not held by qualifying tenants (section 98(1)). 'Approvals' are defined by section 98(7) as including consents and licences and any form of approval to satisfy a restriction registered against the leasehold title at HM Land Registry.

Any rights of approvals belonging to landlords or third parties to the leases relating to flats of qualifying tenants are transferred to the RTM company however (section 98(2)), but in a limited fashion. The limitation is imposed by section 98(4), and works as follows:

- Approvals relating to assignments, sub-lettings, otherwise parting with possession of a flat, charging or mortgaging, structural alterations, improvements, or changes of use may not be granted by the RTM company without 30 days' prior notice to the landlord.
- Approvals under any other heading may not be granted without 14 days' prior notice to the landlord.
- If the landlord objects to the grant within the time-limits above, it may not be made unless the landlord subsequently agrees in writing or the question is determined by the LVT (or on appeal from the LVT) (section 99(1)).
- Objection may only be made if the landlord was entitled to withhold approval before the involvement of the RTM company (section 99(2)). Similarly, an objection may not be made subject to a condition being satisfied unless the landlord was previously entitled to make approval subject to such a condition (section 99(3)).
- The objection is to be made by notice to the RTM company and to the leaseholder concerned and, if the relevant approval is for an act of the sub-tenant, to the sub-tenant (section 99(4)). Notice to the sub-tenant presumably relates only to an existing sub-tenant rather than in a case where approval to sub-letting is sought.
- The application to the LVT may be made by the RTM company, the leaseholder or sub-tenant concerned, or the landlord.

Checklist for landlords

1. Take up the entitlement to join the RTM company. If intending to transfer the freehold to an RMC, push it through before the acquisition date.

2. Consider the implications of RTM for management of premises which contain non-qualifying flats and other units.
3. Look to set up dispute resolution mechanisms.
4. Consider the approach to apportionment of any service charge shortfall.
5. Take an account of the service charge fund as at the acquisition date and pass over any accrued uncommitted balance.

Checklist for RTM companies

1. Ensure alternative funding is available comfortably before the acquisition date.
2. Press the landlord for early transfer of uncommitted service charge balances.
3. Be aware of exactly which management responsibilities are to be inherited on the acquisition date.
4. Consider delegating some management functions to the landlord.
5. Check the leases against the provisions for approvals.

Chapter 7

RTM Company's Powers and Obligations

Having reached the acquisition date successfully, the RTM company will already have had to cover a lot of ground. From the point of view of its overall purpose however, the real challenge is to exercise the Right to Manage in practice once it has been acquired. The measure of the RTM company's success must be its ability to fulfil the management functions (set out in Chapter 6) at least as efficiently and economically as the previous manager.

Furthermore, the challenge has to be accepted from day one. There is no warming-up period. The full burden of the management obligations must be borne from the acquisition date. It is too late then to embark upon strategic planning. The building must be insured, maintenance and repair programmes must be continued or commenced, and it may be necessary to take over immediate supervision of any ongoing repair or improvement projects.

Fulfilling the management responsibilities

Many RTM companies will not have members with direct experience of property management of any sort, let alone the esoteric world of managing blocks of flats. The directors will however have had a period of at least four months between serving the section 79 claim notice and the acquisition date within which to seek the necessary skills or employ a managing agent who already possesses them.

Training in this field is in its infancy, but a variety of literature and lectures are available if the directors know where to look: the columns of the *Estates Gazette* are a good place to start. Depending upon how quickly the Right to Manage is taken up, the likes of managing agents, insurers and lawyers may see providing training to RTM company directors as a potential source of income and, more importantly, future business.

In whatever way education may be found, the areas of knowledge which should be covered ideally include:

- Company law and practice.

- Accounts.
- Landlord and tenant law (with particular reference to service charges).
- Employment (if appropriate).

In any event, RTM companies generally will be encouraged to employ a managing agent who should already have the expertise required to deal with such matters. Apart from the skills factor, there is a strong advantage for delegating day to day management matters to an agent who can be detached from the conflicts of interest which will arise inevitably for RTM directors who are also individual leaseholders. Managing a block of flats with leaseholders with often disparate interests will always require compromising between different priorities. In particular, the RTM company must take a long term view of managing and preserving the property and the leaseholders' funds; individual leaseholders tend to have a shorter term view as their involvement in the property will itself have a shorter term.

Preferably, the choice of whether to employ a managing agent should have been made at a very early stage (indeed, an indication of the company's intentions is required for the Notice Inviting Participation under section 78). The decision could be made later however; the more information the RTM company picks up before the acquisition date the more it may be alerted to the complexities involved.

Managing agents come in a number of shapes and sizes, and might be found in local directories or through organisations such as the Leasehold Advisory Service (LEASE), the Association of Residential Managing Agents (ARMA) or the Association of Retirement Housing Managers (ARHM). The best recommendation, as always, is word of mouth.

Whoever is actually responsible for directly carrying out the management functions of the RTM company, the fundamental decisions will have to be made by the RTM company.

Running the company

The rules and regulations for running the company are set out in the prescribed memorandum and articles of association (see Chapter 1). The decision-making processes are generally standard for most small limited companies, but there are some unique aspects of an RTM company which are worthy of highlighting:

- The quorum required for a general meeting is 20% of the members, or two if that is a greater proportion. In other words, if the company has 10 members or fewer, at least two must be present; if the membership is greater than 10, at least one-fifth must be in attendance for a meeting to proceed. It is therefore conceivable in a small RTM company that one or both members present at a meeting could be landlord members.
- At any time when there are no landlord members of the company (which will certainly be the case until the acquisition date and possibly thereafter) each flat will simply have one vote, to be cast by the qualifying tenant for that flat. Consequently, the votes will not reflect such factors as different sizes of flats or service charge apportionments. It is possible therefore that decisions will be taken by those who have to pay least towards the cost of the company's decisions.
- The votes for each flat must be cast by the qualifying tenant for that flat. Where more than one person constitutes the qualifying tenant (for example husbands and wives or other family arrangements, or tenants in common), any one of them may cast the vote. If more than one attempts to vote however, the chairman may only accept it from the senior person, being the person first named in the register of members. It can be anticipated that joint tenants or tenants in common will not always agree, so the company secretary may have to tread a difficult diplomatic path when entering the members' names on the register.
- If and when there are landlord members of the company, the voting formula is undoubtedly complex. This is dealt with separately below.
- A director need not be a member of the company; it could be the managing agent for example, although it is more usual for agents to serve as company secretary. A director who is not a member has no vote. A director who is a member, and then ceases to be a member (by selling his flat for example) will no longer be eligible to serve.
- The quorum for meetings of directors shall be 50% of the appointed directors, or two, whichever is the greater. The minimum number of directors is two; there is no prescribed maximum, so every company member could be a director.

The voting formula with landlord members

The question of how to arrive at a fair and balanced voting structure for RTM companies with landlord members was one of the most deliberated upon during the consultation process which led to the 2002 Act, and then for the year or so between the passage of the Act and the commencement of the regulations. It was perhaps inevitable that the final outcome would be fraught with complexity.

The full details may be found in Article 39 of the prescribed articles of association, but they may be summarised as follows:

- Each residential unit will receive the same number of votes as the total number of landlord members of the company. Normally, there will be only one landlord, so in most cases each flat will receive one vote.
- Any non-residential part (such as a ground-floor shop or office) will be allocated votes equal to the total votes of the residential units multiplied by a factor of A/B. A = the total internal floor area of the non-residential parts and B = the total internal area of all the residential parts. The method for calculating the floor area is set out in Article 39(b) and paragraph 1(4) of Schedule 6 to the 2002 Act (which deals with the original calculation of percentages to establish whether the extent of the non-residential areas render the premises exempt from RTM – see Chapter 1).
- Clearly the previous paragraph will only apply if there are any non-residential parts in the premises, and they cannot exceed 25% of the internal floor area if RTM is to be exercisable. To attempt a rough example based on the formula so far, if the premises concerned consists of 12 flats on three floors and commercial premises on the ground floor accounting for 25% of the internal floor area, and there is one landlord for the whole of the building, the votes allocated would be as follows:
  - Each of the 12 flats receives one vote, giving a total for the residential units of 12.
  - The commercial area receives 12 votes multiplied by A/B, which will equate to 12 x (25/75) = 4.

  In other words, the commercial areas will receive votes precisely reflecting their proportion of the internal floor area. They will not receive any further allowance in respect of external areas even if they are within their demised premises.

- If there is no member who is a qualifying tenant for a residential unit, the vote is allocated to the member who is the immediate landlord. Accordingly, a landlord with retained or reserved flats will receive an additional vote in respect of such a flat. If however there is no lease for a residential unit, no votes will be allocated in respect of it.
- Votes relating to non-residential parts are allocated to the immediate landlord of that part or, if there is no lease covering it, to the freeholder.

Generally, as with most types of company, the majority of the decisions will be taken by the board of directors, with the members as a whole retaining the key power of appointing and removing members of the board. Each RTM company with a landlord member will need to decide whether it is appropriate to elect a landlord to the board. This will be influenced by the RTM company's relationship with the landlord which, in turn, is likely to reflect the original motivations for exercising the Right to Manage. None the less, given the extent to which RTM companies will need two-way co-operation with the landlord, it is a point which deserves serious consideration.

Of course, in some cases the variety of different hats worn by the landlord (especially in a building with retained or non-residential units) could mean that landlords have a majority of the votes entitled to be cast, so presumably they will vote themselves on to the board.

Enforcement of covenants

A pre-requisite to the RTM company's ability to fulfil its management functions is to have the funds to do so. It is unlikely that the company will have alternative sources of income, except perhaps borrowings which would have to be repaid in due course.

This will be a problem of immense proportions if there were no significant funds passed over from the previous manager (discussed in Chapter 6), which could have occurred for a number of reasons:

- The landlord had not collected advance funding from leaseholders or built up a reserve fund.
- There may be a dispute as to the amount to be handed over which is pending determination by the LVT under section 94(3).

- The leases simply do not allow for payments on account of service charges.
- There may have been substantial arrears of contributions from one or more leaseholders as at the acquisition date.

The power to enforce against leaseholders whose breaches occurred prior to the acquisition date rests with the landlord (implied by section 97(5)). Clearly it would be in the interests of the RTM company for the landlord to enforce service charge arrears promptly in order to preserve and secure the service charge fund. The landlord may see things differently however. Even if he has no desire to be obstructive to the ability of the RTM company to carry out its functions effectively, he may have other reasons for soft-pedalling on arrears recovery. Litigation has never been cheap or free from risk, but the new provisions in the Act aimed at further restricting the use of forfeiture (see Chapter 16) will make the task of recovery even more long-winded and expensive. Even very substantial arrears may become prohibitively costly to collect.

Moreover, if the arrears concerned are of payments in advance of planned future expenditure, landlords may be additionally wary of embarking upon litigation when they no longer have control of how such funds are intended to be spent (that control having passed to the RTM company). Landlords may therefore elect to sit tight and see how things develop, or choose to take action only if indemnified against any shortfall in their costs by the RTM company (which could create a vicious circle of shortage of funds).

From the point of view of leaseholders' breaches of covenant which occur from the acquisition date, these are enforceable by the RTM company by virtue of section 100. Section 100 describes leaseholders' covenants as 'untransferred', because the operation of the Act does not strictly alter the contractual relationships of the parties to a lease when the Right to Manage is acquired. It would be more accurate to say that the existence of an RTM company which has acquired the right has the effect of diverting the parties' existing contractual obligations. Thus, although the covenants are 'untransferred', the principal right to enforce them is transferred on acquisition to the RTM company.

None the less, section 100(2) specifically provides that 'untransferred covenants' are still enforceable by those contractually entitled to do so, as well as by the RTM company. Superficially, this seems at odds with the passing of management powers from the landlord to the RTM company. It must be necessary to deal with matters in this way however, because (a) the landlord may still need to take forfeiture action to ensure performance of covenants (the RTM company has no such powers (section 100(3)), and (b) some covenants are made jointly with other leaseholders as well as the landlord (for example, covenants to observe regulations preventing noise nuisance).

Meanwhile, section 100(5) gives to the RTM company the right to exercise any powers under the lease in favour of the landlord or a third party to enter premises to ascertain whether a tenant is complying with his covenants. Over the centuries, landlords have acquired common law rights to enter upon demised premises in certain situations, over and above powers set out in the lease. There is nothing to suggest that an RTM company also acquires these rights, at least until future case law intervenes.

The general questions of enforcement of covenants will be looked at in more depth in Chapters 16 and 17. For present purposes, suffice it to say that the following methods of enforcement are available to RTM companies:

- Debt recovery by county court proceedings or debt collection services.
- Enforcement of judgments by warrants of execution, charging orders, attachment of earnings orders or third party debt orders (formerly 'garnishee' orders).
- Injunctions restraining breaches of covenant.
- Decrees of specific performance of covenants.
- Damages.

Each of these methods is likely to lead to the RTM company incurring costs, not all of which may be recovered from the leaseholder in default. It will be rare for the RTM company to be able to take advantage of leaseholders' covenants to pay costs incurred as a result of a breach, because the vast majority of such covenants are linked to the forfeiture process. Consequently, enforcement will create an additional cost for the RTM company; whether it can pass that cost on to the leaseholders as a whole through the service charge will depend upon what the lease authorises as service charge expenditure. RTM companies (and those advising them) would be prudent to research what insurance schemes (if any) might be available against such risks. Even so, the RTM company would still have to fund the insurance premium.

As stated above, section 100(3) expressly bars RTM companies from using forfeiture or re-entry. Implicitly therefore it would be unlawful for RTM companies to threaten forfeiture (by service of notices under section 146 of the Law of Property Act 1925 or otherwise). Only a landlord may employ forfeiture, and then only as allowed by the lease or statute. The most that an RTM company may do in this context then is to request the landlord to take forfeiture action.

Monitoring and reporting covenants

Whether or not the RTM company wishes to prevail upon the landlord to exercise forfeiture as a remedy against a defaulting leaseholder, it has a duty to monitor the performance of leaseholders' covenants and report breaches to the landlord (section 101). This provision works as follows:

- The RTM company must 'keep under review' leaseholders' compliance with their covenants (section 101(2)).
- Any breach must be reported to the leaseholders' landlord within three months of the breach coming to the RTM company's attention (section 101(3)).
- However, the breach need not be reported if it has been remedied (within the three months presumably), reasonable compensation has been paid, or the landlord has notified the company that it need not report breaches of that type (section 101(4)).

In the case of compensation, the Act does not state to whom compensation is payable, or who determines what is reasonable. The intention can only be that this is a matter for the RTM company's discretion.

When a breach is reported to the landlord, it will be up to him to decide whether to deploy his enforcement options. Obviously he will have an incentive if the nature of the breach is such that there may be some damage to his reversionary interest: for example, structural alterations without consent. The incentive will be less in respect of breaches ranging from non-payment of service charges to nuisance to other residents.

If the RTM company wishes to persuade the landlord to act, they may be required to proffer security for the landlord's costs and an indemnity. This may be so particularly with RMC landlords. The Act imposes no sanction upon the landlord for declining to respond to a request from the RTM company to exercise his right of forfeiture.

Individual leaseholders may have some contractual rights to call upon the landlord to act (in a nuisance case for example) depending on the wording of the lease; but the lease may also allow the landlord to insist upon security for costs or an indemnity in such circumstances.

To come back to the initial premise of this chapter, the challenge for the RTM company is to carry out its management functions successfully having acquired the Right to Manage, and then to continue to do so. To achieve this, the RTM company will need funds. Assuming that in every block of flats there is at least one leaseholder with a poor payment record (which experience suggests is an accurate assumption), each RTM company runs the risk of falling short on the revenue side. In order to combat the potential for insolvency (and thus failure of RTM) which flows from lack of service charge income, each RTM company will need a plan for how it proposes to deal with arrears. It is probable that straightforward debt recovery procedures will fail to bring in the cash in every case, and so the necessity of obtaining co-operation from the landlord wherever possible is something which RTM companies are likely to face.

Statutory functions

In addition to the management functions set out in various ways in sections 96 to 100, and defined more particularly by the provisions of the leases concerned, Schedule 7 to the Act (introduced by section 102) confirms that many of the statutory liabilities imposed upon landlords will also be borne by RTM companies.

To summarise the statutory functions briefly:

- Section 19, Landlord and Tenant Act 1927: affecting covenants against assignment. Similarly, the Landlord and Tenant Act 1988.
- Section 4, Defective Premises Act 1972: the RTM company will take on the duty of care under this Act from the acquisition date.
- Section 11, Landlord and Tenant Act 1985: the RTM company will take on the duty to keep premises under its management fit for human habitation to the extent that the 1985 Act requires regarding long leases.
- Sections 18 to 30, Landlord and Tenant Act 1985 (except section 26): these are the standard statutory provisions relating to service charges. Section 26 relates to local councils and certain other public authorities.

- Section 30A, Landlord and Tenant Act 1985 and its Schedule: the duty upon the RTM company to give information concerning insurance extends to a landlord as well as leaseholders.
- Section 30B, Landlord and Tenant Act 1985: in the event that there is a recognised tenants' association, the RTM company has a duty to consult it over selection of managing agents.
- Section 5, Landlord and Tenant Act 1987: when a landlord serves an offer notice giving leaseholders the right of first refusal on an intended sale, a copy must also go to any RTM company.
- Part 2, Landlord and Tenant Act 1987: subject to some minor qualifications, the LVT may appoint a manager for premises managed by an RTM company, and the landlord as well as leaseholders may apply for an appointment to be made. The LVT may order further that RTM shall cease to be exercisable for the premises.
- Part 3, Landlord and Tenant Act 1987: where a manager is appointed, the compulsory acquisition of the landlord's interest does not apply.
- Sections 35, 36, 38 and 39, Landlord and Tenant Act 1987: an RTM company is to be considered a party to a lease for the purpose of variation applications.
- Sections 42 to 42B, Landlord and Tenant Act 1987: the RTM company will be a trustee of service charge funds, required to hold them in a separately designated trust account (see Chapter 12).
- Sections 46 to 48, Landlord and Tenant Act 1987: the RTM company must provide information to landlords and leaseholders alike regarding addresses for service and so forth.
- Chapter 5, Part 1, Leasehold Reform, Housing and Urban Development Act 1993: both leaseholders and landlords have a right to a management audit in respect of premises managed by an RTM company.
- Section 84, Housing Act 1996: a recognised tenants' association may appoint a surveyor to advise it on service charge matters in connection with premises managed by an RTM company.
- Schedule 11, Commonhold and Leasehold Reform Act 2002: the new regime affecting administration charges will also apply to RTM companies.

The extent to which these statutes will affect an RTM company is complicated by a variety of modifications and qualifications. RTM company directors and their advisers would do well to make themselves familiar with their consequences. Generally, subject to those modifications and qualifications, RTM companies will face similar rules and strictures to landlords managing leasehold premises.

Non-participating leaseholders

It is worth remembering that the RTM company will owe duties to all leaseholders in respect of its management functions and that, as is the case presently with RMC's, an individual leaseholder cannot be assumed to be in sympathy with the company at all times simply because he is a member of the company. Neither is a leaseholder deprived of his statutory rights by his membership, even if he voted for a decision he later disputes in his capacity as a leaseholder.

Leaseholders who never joined the RTM company, either at the outset or by later admission, or those who resigned their membership will be more likely to fall into dispute with the company. The fact that management is now in the hands of an RTM company made up of fellow leaseholders will not preclude them from taking up cases against the company through the courts or the LVT.

RTM companies should always be aware of the need to keep all leaseholders 'on side', and again the key will be communication. If disagreements cannot be resolved amicably, or if leaseholders perceive the RTM company as failing in its duties, there is a risk that one or more leaseholders may exercise their rights, including:

- An application to the LVT to determine the reasonableness of service charges.
- An application for the appointment of a manager.
- Court proceedings for damages for breach of covenant.
- Simply withholding payment of service charges (see Chapter 12).

The RTM company will be in an especially invidious position should its membership slip below 50% of the qualifying tenants. The only point at which the RTM company had to meet the criterion of having a minimum of 50% of the qualifying tenants on board was when the claim notice under section 79 was served. It is possible that the numbers may have fallen below 50% even by the acquisition date; obviously there is a greater chance of that happening after the acquisition date – especially if unpopular decisions had to be taken. The drop in numbers could simply occur by 'natural wastage', with flats being sold and purchasers not electing to take up membership of the RTM company.

It could well be then that within a very short time after the acquisition date, the premises could be managed by a small minority of the leaseholders, perhaps with an even smaller minority engaged actively.

In such circumstances, the probability of any of the applications listed above being made will grow exponentially.

The position of the landlord

In terms of the duties owed by the RTM company, the landlord (whether an investor freeholder or an RMC) takes up a position very similar to that of the leaseholders, and with similar remedies. The landlord's priorities will differ somewhat from those of the leaseholders, but they will have a common interest in the property being effectively managed, repaired, maintained and insured.

Accordingly, in the event of the RTM company failing to meet its obligations (or possibly even appearing to do so), it could face LVT applications or court actions from the landlord as well as from leaseholders. It is perhaps more likely that landlords will be willing to meet the costs of such applications, and they will not be concerned about the prospect of suing a neighbour.

Checklist for RTM companies

1. Have planned maintenance and other programmes ready for implementation on the acquisition date.
2. Company directors should take advantage of any available training.
3. Reconsider appointing managing agents, if not already done.
4. Develop a planned approach to debt recovery.
5. Look around for insurance cover for shortfalls on the costs of recovering arrears.
6. Prepare a procedure for monitoring and reporting breaches of covenant.

7. Be prepared to seek co-operation from the landlord when forfeiture might be required.
8. Be aware of the management functions required by statute as well as by the leases.
9. At all times, keep all the leaseholders on board.

Chapter 8

Cessation of RTM

There are a number of circumstances in which the exercise of the Right to Manage might come to an end. Some events may be entirely voluntary, and others imposed. In Chapter 4 we looked at the ways in which a claim to RTM can be withdrawn before the right is acquired (and the consequences of withdrawal). In this chapter we examine how the events leading to cessation of RTM after acquisition could come about.

Cessation events

Section 105 of the 2002 Act sets out the circumstances by which an RTM company may be rendered unable to continue to exercise the right of management:

- An agreement may be made between the RTM company and the landlord (or, in the case of a variety of different landlords for all or part of the premises, each landlord).
- A winding-up order or administration order is made against the company.
- The company passes a resolution for voluntary winding-up.
- A receiver or manager is appointed with regard to the RTM company's undertaking.
- A debenture holder in respect of a floating charge secured over any of the company's property takes possession of that property.
- There is approval of a voluntary arrangement regarding the company under Part 1 of the Insolvency Act 1986.
- The RTM company's name is struck off the register under sections 652 or 652A of the Companies Act 1985.
- A manager is appointed by the court or LVT in relation to the premises under Part 2 of the Landlord and Tenant Act 1987 (section 105(4)).
- The company ceases to be an RTM company in relation to the premises (for example, if it acquires the freehold title).

The first and last of these events suggest that RTM may cease because the leaseholders concerned have moved on, perhaps to a stronger position, such as by purchasing the freehold. In such a case, the exercise of the Right to Manage will have become unnecessary. Where an RTM company becomes the freeholder, it should be relatively simple to carry out the requisite changes to the company's memorandum and articles of association.

In due course, the premises may again be the subject of a claim notice for a new attempt at RTM, but generally not for at least four years after the conclusion of the original exercise (paragraph 5, Schedule 6 to the 2002 Act). This is not an absolute bar however. The exemption will not apply at all if RTM ceased simply because the RTM company acquired the freehold. Furthermore, the LVT may determine that it is unreasonable to exclude RTM in particular circumstances (paragraph 5(3), Schedule 6).

The remaining cessation events are more likely to occur if there has been some fault or failing on the part of the RTM company. These can be conveniently assembled into three categories: insolvency; non-compliance with company rules; mismanagement.

Insolvency

Insolvency may be the greatest threat to RTM companies, and could affect even the most efficient and best intentioned. A company will be insolvent if it is unable to pay its debts. An RTM company can only pay its debts (principally the cost of performing its management functions) if it can raise sufficient income through service charges.

It is not only recalcitrant leaseholders who create shortfalls in service charge recovery. It may be that there are substantial arrears still owed to the landlord (and therefore the service charge fund), which the landlord may be in no hurry to collect. Especially (but not exclusively) in older properties, there may be weaknesses in the leases which preclude managers from making full recovery of certain expenditure items.

It could also be that the LVT has reduced a service charge bill on a reasonableness application, or because of some technical breach: for example of the new consultation requirements in the 2002 Act (see Chapter 11).

It is very unlikely that the RTM company will have any capital assets to help it survive a shortage in funds. Its only resource apart from the service charge is the pockets of its members. How deep those pockets are will obviously vary from case to case.

Because of the contrast between the immediacy with which management responsibilities pass to the RTM company and the inevitable delays caused by the inability to raise service charges on demand, the risks of insolvency facing an RTM company can arise instantly on the acquisition date. This is the fundamental reason for taking measures well before acquisition to put the RTM company in funds to carry out its essential obligations.

Non-compliance with company rules

The formalities required of an RTM company are relatively slight (compared to a major corporation), but they can be daunting none the less for directors with no previous experience of running a company. It can be foreseen that few flat-owners will have such experience.

Most managing agents should have the ability to offer company secretarial services, or at least to advise upon the relevant duties. When considering the appointment of managing agents (which should be done at a very early stage in the process towards exercising RTM), this is one of the factors to be taken into account.

Mismanagement

An RTM company does not have to show fault on the part of the existing manager to be able to exercise the Right to Manage. It is not itself open to such an attack. However, the Right to Manage is not a licence to mismanage, and an RTM company can be the subject of an application by individual leaseholders or landlords to have management taken away in favour of a manager appointed by the LVT.

The grounds for appointing a manager are set out in full in Part 2 of the Landlord and Tenant Act 1987, but they include making unreasonable service charge demands, failing to perform management obligations, and non-compliance with approved codes of practice (such as the RICS management code). It is perfectly feasible that an RTM company could be held to have acted in such a way, perhaps entirely innocently. Indeed, its failures could have been caused by factors outside its control (non-payment of service charges for instance).

The LVT has a discretion on these applications, but it is by no means beyond the realms of possibility that an RTM company could find a Manager appointed. The Manager would take over the company's management functions, and the Right to Manage would be lost.

An RTM company which holds the sympathy of all leaseholders (whether or not they are members of the company) will be less vulnerable to such an application. Communication is a vital factor in retaining leaseholders' support and understanding, but it is also crucial that the RTM company directors comprehend exactly what is required of them, so it is advisable for them to gain familiarity with the relevant codes of practice and the extent of their duties in the context of the leases for the premises.

What happens after cessation of RTM?

The 2002 Act does not attempt to describe the position following the cessation of RTM. To a certain extent, what happens next will be determined by the manner in which the Right to Manage ceased to be exercisable.

- If the cessation event was an agreement between the RTM company and the landlord, then it is probable that what follows will have formed part of the agreement. If not, then the route must be the same as an enforced cessation (for example, in the circumstances of insolvency – see below).
- If RTM came to an end simply because the RTM company became the freeholder, then the company in its new form would continue to be responsible for management of the premises.
- If a manager is appointed under the 1987 Act, then it is the manager who takes over the management functions (albeit in the shoes of the landlord).
- In any of the other cases, there will not be either a contractual or a statutory provision to deal with ongoing management.

Logically, the exercise of the Right to Manage must be seen as interruption to the landlord's management obligations. The interruption will be made permanent if the RTM company acquires the freehold subsequently, but in any other circumstances it can only be temporary. Consequently, if the RTM ceases to be exercisable by any means other than transfer of the freehold title, it must be assumed that the previous state of affairs will apply and that the landlord will resume his position as manager (or, depending on the precise situation affecting the premises, some other party such as a residents' management company).

Of course, it could not possibly be that things will be entirely as before. The RTM company will have changed matters, so it would be impossible for the landlord to take over as though RTM had never happened. Indeed, it may well be that the landlord would not want to take management back. Nevertheless, if it is right that he will be bound by his covenants once again, he will have little choice.

The position would be more serious if the previously ousted manager had undergone a fundamental change since RTM was acquired. If the manager were an RMC for example, it could easily have ceased to exist (given that its entire *raison d'etre* was the management of the premises). In such circumstances, the previous manager would be incapable of resuming management, leaving a vacuum. The leases might provide for this by requiring the freeholder to step in if the RMC fails (which could leave the leaseholders in a worse position than they had occupied prior to RTM).

If there is no party in a position to step in after cessation of RTM, the leaseholders will be in a very difficult situation (as will the freeholder, if one exists in this scenario).

There are a number of alternative solutions, depending upon the facts affecting the individual property. They include:

- An application to the Companies Court for the restoration of the residents' management company.
- An application to the LVT for the appointment of a manager.
- In the case of an absentee or non-existent freeholder, an enfranchisement application.
- The formation of a new RTM company (which would only be possible if the LVT permits it).

Most of the actions which could be taken will have to be pursued by leaseholders, the most active of whom were probably directors of the RTM company. Whether the RTM directors could muster sufficient support from their fellow leaseholders must be open to question – especially if the RTM company represented only a bare majority (if that) of the leaseholders when it was in being.

What is certain is that any of the steps which could be taken would be expensive for leaseholders who may already be under considerable financial pressure at this stage. They would also take time. In the meantime, the management of the premises will be at minimum levels at best, putting at risk such vital services as insurance and repairs.

Even if there is a landlord or other manager ready to step in when RTM ceases, the consequences will be less than straightforward.

If the previous manager's agreements with contractors were frustrated by the transfer of management to the RTM company, then logically the same must happen with the RTM company's contracts – and potentially with considerably shorter notice. The manager will almost have to start again. He will of course be required to go through all the consultation procedures in respect of new contracts, or apply to the LVT for dispensation from consultation.

Meanwhile, the accounts will have to be sorted out (which may include tracing funds); insurance must be placed if it is not still on foot; essential maintenance and repairs will need to be carried out.

The list of additional work to be done by the manager will be long and involved.

The hiatus caused by the cessation of RTM will also have significant consequences for leaseholders. For example:

- The manager will expect leaseholders to meet the considerable management costs of putting things into order, so the service charge will increase.
- For a period, possibly a long period, services will be interrupted.
- Flats will be very difficult to sell or mortgage during any period of uncertainty over management of the building.
- Sale prices may diminish.
- From a psychological point of view, leaseholders may feel defeated and deflated. They may regard the balance of power as having swung back decisively in the landlord's favour.

All in all, the experience could leave leaseholders feeling that the exercise of RTM had left them worse off than they were before the idea was first raised. Any resultant bitterness would hardly benefit landlords either; the atmosphere could render the premises very difficult to manage, with one possible upshot being the return of the days of high levels of litigation over service charges.

There may even be additional concerns for leaseholders who were directors of the RTM company, especially if the company was insolvent. Directors who have permitted a company to continue to trade despite possessing the facts to know that the company was unable to pay its debts can be held to have indulged in 'wrongful trading'. Such a finding can result in directors being disqualified from future directorships (a powerful sanction in itself) and possibly pursued personally for the company's debts.

This may be a fairly remote contingency, but it is feasible, and the risk is among those for which RTM directors should be seeking insurance cover if it can be found.

Conclusions

None of this should be seen as an argument against attempting RTM as such. The range of potential consequences of the exercise failing is, however, a compelling reason for taking a serious and cautious approach to any decision about whether to tackle it.

No decision regarding whether to embark upon the RTM path should be taken in isolation from an examination of the alternatives. For example:

- RTM also presents risks for a landlord. At the very least, the process is likely to involve discomfiture and expense. It is possible therefore that the threat of RTM will be sufficient to open negotiations towards improvements in management issues.
- Purchase of the freehold may require the expenditure of substantial sums in terms of price and costs, which obviously can be off-putting. Any calculation of this should take into account the potential costs which could be incurred throughout the RTM process, many of which are highlighted in this volume.
- Leaseholders have other statutory rights and remedies which might achieve similar results. Many of these rights are strengthened and extended elsewhere in the 2002 Act. In particular, applications to the LVT on a wide range of service charge matters should not be overlooked. Additionally, leaseholders' right to a management audit under the Leasehold Reform Housing and Urban Development Act 1993 (as amended by Schedule 10 to the 2002 Act) is not used as often as it might be.
- If the motivation for exercising the Right to Manage springs largely from unrest with the landlord's management, whether in terms of cost, standard or adequacy, it may be as well simply to apply to the LVT for the appointment of a manager (which may itself lead to the freehold passing to the leaseholders in due course).

Indeed, one of the very first points to consider when reviewing the options is the reason for thinking about RTM in the first place. Different leaseholders may have different reasons. Being clear about the long term objective is essential to selecting the best option.

If RTM is to be the chosen path, it is important that the directors of an RTM company, and indeed its members and potential members, should understand the variety of possible circumstances under which the Right to Manage can cease, and the consequences of cessation. Without this fore-knowledge, it will be the more difficult to guard against those eventualities from the outset.

Particularly given the amount of time which must pass between the first stirrings of interest in exercising RTM and the eventual acquisition date (at least four or five months and possibly over a year), it would be prudent to have in mind at every stage what can go wrong and the likely consequences.

This is especially so where there is a likelihood of personnel changing over the development period (through flats changing hands, fluctuating levels of interest and so forth).

RTM company directors face an exacting task for which they need to be well-prepared. It is hoped that this book will assist in the process of preparation, but every opportunity for information and training should be grasped.

Chapter 9

Collective Enfranchisement

Although it is not the intention of this book to give a detailed analysis of the new enfranchisement provisions in the 2002 Act (as mentioned in the Introduction), it is relevant to look at the subject briefly in the context of leaseholders acquiring the management of their buildings – not least because it is one of the alternatives to electing to exercise the Right to Manage. Enfranchisement is also one of the events leading to the cessation of RTM.

As outlined in the previous chapters, one of the principal disadvantages to the Right to Manage is that an RTM company has no power to threaten the ultimate sanction to a defaulting leaseholder – forfeiture. This is because the RTM company does not acquire the freehold title to the premises. Moreover, RTM will often be a temporary respite from management by the landlord; in the majority of circumstances by which RTM may fail, the right to manage will revert to the landlord.

Clearly this will not be the case if the leaseholders acquire the freehold, by enfranchisement under the statutory provisions or simply by negotiation.

The Right to Enfranchise

The amendments to the Leasehold Reform, Housing and Urban Development Act 1993 (the 1993 Act) brought in by sections 114 to 128 of the 2002 Act are designed to make the process of enfranchisement simpler and more widely available. Unfortunately, the commencement timetable in relation to these sections has been long and convoluted. Some sections came into force on 26 July 2002; some had not been implemented by the date of publication of this volume. Readers looking to take up any of the enfranchisement provisions under either the 1993 Act or the 2002 Act are advised to check exactly which sections are in force at the appropriate time.

Generally, the intention of the 2002 Act is to make the Right to Enfranchise (RTE) and the Right to Manage as similar as possible in their operation, and particularly in the context of formalities.

Consequently, the Right to Enfranchise will be restricted to a prescribed form of RTE company which will look very much like an RTM company. It is undoubtedly envisaged that an RTM company will mutate into an RTE company, and the Act makes it very easy to do so.

The qualification criteria are also very similar, but of course they must be met by the RTE company when it exercises the right to enfranchise, whereas the RTM company may have fallen from the criteria for RTM by the time its members consider moving to collective enfranchisement.

Acquisition of the freehold by negotiation

Despite the extensive statutory provisions contained in the 1993 Act, and indeed previous legislation conferring a right of pre-emption on leaseholders in the event of a transfer of the freehold, it appears to have been far more common for leaseholders to have acquired their freeholds simply by negotiation with their landlords, entirely outside the statutory machinery.

It is thought that that is unlikely to change significantly, notwithstanding the relatively easier procedures contained in the 2002 Act.

Many freeholders will be quite willing to enter into negotiations for disposing of a single freehold title without the fuss and expense of the valuation process through the Leasehold Valuation Tribunal. In a sense, omitting the statutory enfranchisement procedures may be even more attractive because of the 2002 Act, simply because it will be much easier to determine the price which would have been fixed by the LVT under the 2002 Act's valuation formulae.

Even freeholders who are not initially minded to discuss terms may be brought to the negotiating table by a number of factors, for example:

- The Right to Manage has already been exercised, or is threatened.
- There is a substantial dispute with leaseholders which might lead to lengthy and expensive litigation through the courts or LVT.
- The leases are regarded as defective (by failing to provide for full recovery of service charges for example).
- They receive an offer in excess of what they could expect to achieve through forced enfranchisement.

- The leaseholders represent such a 'nuisance factor' that the property has become uneconomic to manage.

There are distinct advantages to the leaseholders in acquiring their freehold by negotiation:

- The enfranchisement procedure is lengthy and expensive; leaseholders will have to bear their valuers' and solicitors' fees, and quite probably the landlord's professional costs as well. The savings in costs could be used perhaps to offer a slightly higher price for the freehold in order to achieve a speedy transaction.
- The purchasing vehicle need not be tied to the prescribed format of an RTE company.
- The qualification criteria for enfranchisement need not be met (although all qualifying leaseholders will effectively have to be invited to participate through the operation of the right of pre-emption in the Landlord and Tenant Act 1987).
- The precise terms of what is or is not to be conveyed can be more flexible than under formal enfranchisement.

Of course, the resultant freeholder will not be exempt from a subsequent exercise of the Right to Enfranchise (or Right to Manage) by qualifying leaseholders.

The rights and obligations of freeholders

By whatever means leaseholders employ to acquire their freehold, the end result will effectively be the same, and it is necessary to examine this position as part and parcel of the process of deciding whether or not to follow this path. In doing so, it is useful to compare the relative situations of freehold ownership and leaseholders exercising the Right to Manage.

As freeholders, leaseholders will be able to use the right of forfeiture against other recalcitrant lessees; an RTM company cannot. A freeholder is free to grant variations and extensions of leases; generally, an RTM company cannot (although it can apply to the LVT for variations in limited circumstances). A freeholder can reduce the ground rents; an RTM company cannot. A freeholder holds the property in perpetuity (assuming it survives); an RTM company's position is much more transient, and it is potentially open to interference by the landlord. A freeholder has the final say on such matters as consents and licences (subject always to the courts or LVT); an RTM company has a say, but not the final word. A freeholder owns an asset which can be used as security for borrowing; an RTM company does not. A freeholder has an income stream from the rents which can be offset against the company's running costs; an RTM company does not. Leaseholders who elect to go straight for their freehold rather than exercising RTM will save all the costs of RTM (which are likely to be substantial).

Conversely, although both freeholders and RTM companies are subject to a long list of obligations, both contractual and statutory, freeholders are burdened by some matters which do not affect RTM companies. For example:

- The right of forfeiture is severely restricted in its use and effectiveness by decades of legislation, including new restraints introduced by the 2002 Act (see Chapter 16).
- The freeholder owes duties to all its leaseholders and tenants (whether or not they are qualifying tenants for the purposes of RTM), and to neighbouring landowners and occupiers.
- A freeholder may be subject not only to RTM and a variety of LVT applications, but also future attempts at enfranchisement and a whole host of other contractual and statutory remedies.
- If an RTM company becomes insolvent or fails in other ways, it is likeliest that the freeholder or an LVT-appointed manager will take over management. If a leaseholder company owning the freehold collapses, there is no party who can automatically step in and manage the property, with consequential effects on preservation of assets and saleability of flats.

Conclusion

Leaseholders seeking to take over the management of their buildings, or to take control of how it is managed and by whom have two principal alternatives facing them: exercising the Right to Manage or acquiring the freehold.

In the end, the alternatives are not necessarily mutually exclusive, as it is possible for an RTM company to acquire the freehold subsequently, but there would seem to be little point in exercising RTM if the eventual intention is to purchase the freehold title anyway.

Both alternatives involve major decisions and considerable expense. Either will lead to fundamental consequences for the future of the property and all leaseholders and their subsequent purchasers. (There will also be significant consequences for the landlord, and his likely reactions will be a factor to be taken into account when developing the leaseholders' strategy.)

The decision concerning which alternative to explore is obviously one which deserves the most serious deliberation by all concerned, and ideally a fully informed and considered decision should be taken at the earliest possible stage (preferably well before any formal notices are served upon the landlord). Leaseholders will be prudent to avail themselves of every realistic opportunity for taking advice from professionals well versed in the field before a decision is made – let alone at the various steps along the way.

PART 2

SERVICE CHARGES

At the time of going to publication, not all of the provisions in the 2002 Act concerning service charges had come into force. This part of the text will examine the Act as it stands subject (where appropriate) to the secondary legislation which will eventually implement it, and its likely effects in practice. The amount of secondary legislation required will vary considerably from section to section; accordingly the degree of speculation about the Act's consequences will inevitably reflect that.

It is necessary to try to cover the impending service charge changes none the less, particularly so that those considering or actively exercising the Right to Manage will have some idea of what faces them in the short to medium term. Obviously, anyone affected by the changes to the legislation should first ascertain which of the new provisions are in force.

Chapter 10

Widening the Definition

The first fundamental change to service charge legislation introduced by the 2002 Act is to extend the meaning of 'service charge' itself. The alteration is effected by section 150 which amends the definition within section 18 of the Landlord and Tenant Act 1985 (the cornerstone of residential service charge law).

Improvements

Section 18 of the 1985 Act (as amended) will now read as follows (with the insert highlighted by italics):

> (1) In the following provisions of this Act 'service charge' means an amount payable by a tenant of a dwelling as part of or in addition to the rent –
>
> (a) which is payable, directly or indirectly, for services, repairs, maintenance, *improvements* or insurance or the landlord's costs of management, and
>
> (b) the whole or part of which varies or may vary according to the relevant costs.
>
> (2) The relevant costs are the costs or estimated costs incurred or to be incurred by or on behalf of the landlord, or a superior landlord, in connection with the matters for which the service charge is payable.
>
> (3) For this purpose –
>
> (a) 'costs' includes overheads, and
>
> (b) costs are relevant costs in relation to a service charge whether they are incurred, or to be incurred, in the period for which the service charge is payable or in an earlier or later period.

Thus, by the insertion of one word (and a comma), Parliament has settled a long-standing anomaly. The cost of improvements will now be a service charge, and they will now be subject to the full panoply of consultation in the same way as any other service charge item (see Chapter 11). Furthermore, those long-winded esoteric arguments, beloved of property lawyers, as to whether specific works constitute repairs or improvements will be of far less importance.

There is an additional consequence of converting improvements to a service charge item: that is that leaseholders will be able to apply to the LVT for a determination that the cost of improvements or proposed improvements is unreasonable, or for the appointment of a manager (under Part 2 of the 1987 Act) on the grounds that unreasonable improvements have been carried out or are proposed.

Since 1985 (and before) landlords have been accustomed to treating works of improvement in an entirely different way from works of repair. Inevitably, a certain complacency has crept in. Landlords can afford to be blasé no longer in respect of such matters.

Definitions in a lease

It is probably worth pointing out at this stage that the statutory definition of service charges contained in section 18 of the 1985 Act will override any individual definitions in a lease. A landlord cannot escape the constraints of the service charge legislation by relying on a provision in a lease for example that insurance is dealt with separately from the service charge. The insurance must still be in a reasonable amount, and expenditure on it must be supported by adequate documentation made available on request. The same applies to any other form of expenditure to which a leaseholder is expected to contribute so long as it is listed in the heads of expenditure referred to as 'relevant costs' in section 18.

Conversely, just because a head of expenditure is mentioned in section 18, it does not permit the landlord (or any other manager such as an RTM company) to seek to recover his costs through the service charge if it is not specified in the lease at any point. The definition in section 18 may override the definitions in the lease, but it does not supplant them. For example, a lease which fails to provide for the recovery of the costs of improvements will still prevent a landlord from enforcing a contribution from leaseholders. The same would apply to such matters as management charges or legal costs (common causes of litigation).

Administration charges

Although the 2002 Act does not expressly add 'administration charges' to the definition of service charges in section 18 of the 1985 Act, it produces a similar effect through section 158 which introduces Schedule 11 of the 2002 Act. Previously, administration charges were not caught by the service charge legislation. This omission was perceived to provide less scrupulous landlords with an opportunity to make excessive charges for certain services without having to have regard to their reasonableness, and without any scrutiny by the LVT.

Schedule 11 brings in a regime for such charges which is largely parallel to that governing service charges. Indeed the definition of administration charges in Schedule 11 follows the wording of section 18 very closely:

> 1(1)In this Part of this Schedule 'administration charge' means an amount payable by a tenant of a dwelling as part of or in addition to the rent which is payable, directly or indirectly –
>
> (a) for or in connection with the grant of approvals under his lease, or applications for such approvals,
>
> (b) for or in connection with the provision of information or documents by or on behalf of the landlord or a person who is party to his lease otherwise than as landlord or tenant,
>
> (c) in respect of a failure by the tenant to make a payment by the due date to the landlord or a person who is party to his lease otherwise than as landlord or tenant, or
>
> (d) in connection with a breach (or alleged breach) of a covenant or condition in his lease.

By restricting the definition to charges payable 'as part of or in addition to the rent' the remainder of the Schedule would seem to apply only to payments contractually due under the lease. Consequently, some charges would appear to escape, most obviously the fees raised by managing agents for responding to solicitors' conveyancing enquiries. These are not payable under the lease, but rather under a separate one-off contract with the leaseholders' solicitors (notwithstanding that it will be the leaseholder who pays in due course). If the enquiries are made by an intending purchaser who has yet to acquire the lease, there is no landlord and tenant relationship between the parties at all.

It is not yet clear whether subparagraphs (c) and (d) might be construed to include such things as legal costs in enforcement actions. There will usually be contractual provision in the lease to pay the costs (including solicitors' and surveyors' fees) for such matters, but legal costs are also subject to assessment (formerly 'taxation') by the courts (and by the LVT to a certain extent – see Chapter 13) and it may be found to be more appropriate to deal with costs separately.

It is interesting to note however that Parliament has expressly reserved to the appropriate Secretary of State to amend (and therefore extend) the definition of administration charges (paragraph 1(4) of Schedule 11).

The Schedule then goes on to deal differently with administration charges which are fixed in an amount or by a formula specified in the lease and those which are variable (i.e. not fixed by the lease).

Variable administration charges

An example of a variable charge would be a landlord's fee for registering a notice of assignment which the lease simply reserves as 'a reasonable fee'.

Henceforth a variable administration charge will be payable 'only to the extent that the amount of the charge is reasonable' (paragraph 2, Schedule 11). Determinations of reasonableness may be made by the Leasehold Valuation Tribunal.

Fixed administration charges

Paragraph 3 of Schedule 11 gives the LVT the power to vary a lease, on application by any party to it, on the grounds that an administration charge is unreasonable or that a formula for ascertaining the charge is unreasonable. (The LVT is also granted wider powers to make variations elsewhere in the 2002 Act; see Chapter 13.) The application should be accompanied by the draft variation, although the tribunal has discretion as to the final wording. Alternatively, the tribunal may direct the parties to vary the lease in such a way as it specifies.

The paragraph goes on to state (in subparagraph 6) that any variation will be binding 'not only on the parties to the lease for the time being but also on other persons (including any predecessors in title), whether or not they were parties to the proceedings in which the order was made'. A variation to a lease, properly executed, will be binding on successors in title in any event of course. It would be unusual however for a variation to bind predecessors in title, since this would seem to give the variation retrospective effect, and may have consequences for parties who have neither any say in the proceedings nor any continuing interest in the property.

The import of subparagraph 6 is difficult to divine unless and until it has been construed judicially. It may be that leaseholders will try to claw back monies paid to previous landlords before the variation, on the grounds that the charges concerned have been declared as unreasonable by the LVT. Landlords facing such claims are likely to argue that it is unjust to penalise them retrospectively, especially when they were not parties to the proceedings which made the finding of unreasonableness. In any event, the relevant commencement order[1] provides that this paragraph will not apply to an administration charge payable before 30 September 2003. None the less, it is unclear why this subparagraph was introduced unless such claims were envisaged.

Liability for administration charges

Any demand for an administration charge must be accompanied by a summary of leaseholders' rights and obligations relating to such charges (paragraph 4, Schedule 11). A prescribed form for these summaries may be introduced in due course (and indeed this has already been the subject of consultation). None the less, the statutory requirement to serve a summary came into force on 30 September 2003, leaving property managers to invent their own forms for the time being.

If no accompanying summary has been served, a leaseholder may withhold payment, and none of the penalties in the lease for late or non-payment will apply (paragraph 4(3) and (4), Schedule 11).

This sanction against landlords is continued throughout the service charge provisions of the 2002 Act and will be examined more fully in Chapter 12.

Furthermore, a leaseholder may now challenge an administration charge by application to the LVT (whether or not it has already been paid) (paragraph 5, Schedule 11), and the tribunal's determination may go beyond the simple issue of reasonableness to establish whether it is payable (and therefore whether it is a valid charge under the lease), and then:

- By whom it is payable.
- To whom it is payable.
- The amount due.
- The date and manner in which it is payable.

However, the LVT's jurisdiction is excluded if the matter has already been agreed or admitted by the leaseholder, or if the issue has already been determined by the court or arbitration, or if it is has been or is to be referred to a post-dispute arbitration agreement.

No agreement or admission is to be inferred by payment (paragraph 5(5), Schedule 11). If they are asserted in other

[1] Paragraph 8, Schedule 2, Commonhold and Leasehold Reform Act 2002 (Commencement No. 2 and Savings) Order 2003.

circumstances, the LVT will presumably have to determine this as a preliminary issue, which will require evidence of a genuine agreement or admission.

The only form of arbitration which can prevent the LVT from having jurisdiction is a 'post-dispute arbitration agreement': that is an agreement between the parties to refer the question to arbitration once the existence of the dispute is known. An arbitration clause in the lease will not be effective.

There are additional sanctions upon a landlord for demanding an unreasonable charge, by way of amendments to the Landlord and Tenant Act 1987 introduced by Part 2 of Schedule 11 to the 2002 Act. These fall into two main categories:

- Unreasonable administration charges will constitute a further ground upon which leaseholders may apply to the LVT under section 24 of the 1987 Act for the appointment of a manager.
- Administration charges will not be payable unless the landlord has furnished the leaseholder with the information concerning the landlord's name and address required by sections 47 and 48 of the 1987 Act.

Summary

In extending the definition of service charges to include improvements, and treating administration charges in much the same way as service charges, Parliament has brought most conceivable payments which a leaseholder may have to make to a landlord (except rent) into a unified regime. Leaseholders will thus have a much more general protection against unreasonable or unjustifiable demands from landlords.

Landlords will have to review their procedures and charging structures to ensure that they are charging reasonably and within the terms of their leases. Larger landlords and managing agents with substantial portfolios, who have tended to adopt a broad brush approach to such questions notwithstanding the variety of leases under their management, may have to undertake a painstaking trawl through those leases to confirm that they are not seeking to recover monies outside their provisions.

It should be remembered that 'landlords' for these purposes will also include residents' management companies and RTE and RTM companies.

Chapter 11

The New Consultation Regime

The 2002 Act makes fundamental changes to the duties on landlords and managers to consult long leaseholders when proposing substantial expenditure to which leaseholders will contribute through the service charge. These changes came into force on 31 October 2003 by virtue of The Service Charges (Consultation Requirements) Regulations 2003 ('the regulations'). Although contained in section 151 of the 2002 Act, the cornerstone of the consultation legislation remains section 20 of the Landlord and Tenant Act 1985 (as amended by the 2002 Act). Consequently, property managers will still know it as 'section 20 consultation', but they would be very unwise not to notice that the statutory provisions have been modified and extended very significantly.

In fact, section 151 introduces a complete substitution for the main body of section 20 of the 1985 Act, but not for section 20A (grant-aided works), section 20B (time-limits for demands), or section 20C (costs of proceedings). Section 151 also brings in a new section 20ZA. Such piecemeal adjustments to legislation are bound to confuse. Property managers may consider it advisable to put together their own packs assembling the various elements of section 20 which are now in force, and excluding what has gone by the board.

The ethos of section 20 remains the limitation of service charges payable by leaseholders (where expenditure meets the criteria of the section) to a threshold imposed by statute. The regulations under the section fix minimum monetary scales, and where expenditure on relevant items exceeds those scales, the landlord or manager may not recover through the service charge anything over and above the minimum threshold unless the leaseholders have received prior consultation in accordance with the Act.

Section 20 applied previously only to expenditure on specific works to a property. A simple example of its application would be if the cost of works was greater than £50 per contributing leaseholder or £1,000 for the leaseholders together, the landlord could recover no more than £50 per flat or £1,000 in total if he had not carried out prior consultation, even if the cost was vastly over those limits. There were a few exceptions, but that was the general rule, and it could not be evaded even if the cost was otherwise entirely reasonable.

Qualifying agreements

The principal change brought in by section 151 is not to that basic philosophy, but to extend the application of it to a hugely increased range of headings of expenditure. The headings are sorted into two categories: 'qualifying long term agreements'; and 'qualifying works'. A qualifying long term agreement is any agreement (subject to exceptions in the Regulations) entered into by the landlord or a superior landlord which relates to service charge matters for a term of more than 12 months (section 20ZA(2)). Qualifying works are 'works on a building or any other premises' (section 20ZA(2)).

Originally, in the consultation process on the Bill which became the 2002 Act, it was proposed that qualifying long term agreements would be for 12 months or more, which would have caught such things as insurance renewals. Sensibly, this was adjusted, so no contracts for a short term of 12 months or less will now be affected by the consultation requirements. Of course, they must still be reasonable (section 18, Landlord and Tenant Act 1985), but managers are saved from an inordinate amount of consultation on such agreements.

The Act helpfully clarifies a long-standing dilemma by stating that the consultation requirements also apply to agreements entered into by superior landlords (the Regulations clarify still further that this affects equally long term agreements and specific works).

Definition of the 'landlord'

It is important to note that the definition of 'landlord' for the purposes of section 20 will remain that set out in section 30 of the 1985 Act: namely 'any person who has a right to enforce payment of a service charge'. Consequently, this must include:

- Freeholders.
- Resident Management Companies ('RMC's').
- Right to Enfranchise ('RTE') companies.
- Right to Manage ('RTM') companies.

If there is an RTM company or an RMC (which does not own the freehold) it is they who will be the 'landlord' in the context of this part of the legislation rather than the freeholder. It would have been less confusing perhaps had the Act referred to a 'manager' rather than a 'landlord', but it is essential that RMC's and RTM companies understand this distinction.

Exceptions and dispensations

The possible dispensation from the consultation requirements is now brought up-front. Section 20(1) now provides that the limitation will apply to all qualifying works or agreements unless there has been compliance with the consultation rules or an LVT (or an appeal from the LVT) has determined that they can be dispensed with for the specific works or agreement. The tribunal may so determine 'if satisfied that it is reasonable to dispense with the requirements' (section 20ZA(1)).

Otherwise, the meat of the new section 20 is contained in the Regulations.

The following are excluded from the definition of qualifying long term agreements by paragraph 3 of the Regulations:

- Contracts of employment.
- Management agreements made by a local housing authority and a tenant management organisation ('TMO') or a body established under section 2 of the Local Government Act 2000.
- An agreement between a holding company and its subsidiary, or between subsidiaries of the same holding company (the definitions following those in the Companies Act 1985).
- An agreement for less than five years which was entered into at a point when there were no tenants or leaseholders at the property (although it is not apparent how consultation could take place without leaseholders to consult anyway, however long the duration of the agreement).
- An agreement exceeding 12 months which was entered into before 31 October 2003 (even if there are more than 12 months of the contractual term to go).
- An agreement for qualifying works in respect of which public notice was given before 31 October 2003.

Apart from the exclusions listed above, the Act contains no further general exemptions, so qualifying long term agreements which become subject to the consultation requirements (so long as they exceed 12 months) could include any or all of the following:

- Managing agents' agreements.

- Maintenance agreements for such things as lifts or entry-phone systems.
- Cleaning and gardening.
- Utilities.
- Possibly even retainers for surveyors, accountants and solicitors.

It is foreseeable that several of these services might be provided realistically by only one supplier. None the less, if it is proposed to enter into an agreement for over 12 months, either consultation must be carried out or dispensation sought from the LVT.

The consultation requirements for qualifying long term agreements

The section 20 consultation requirements will apply to a long term agreement if the amount payable towards it by any tenant (note: not each tenant) will exceed £100 in any accounting period (ie 12 months). There is a somewhat complicated formula for ascertaining the precise accounting period in paragraph 4 of the Regulations, but generally this will start to apply to individual properties from the start of the next complete accounting period after 31 October 2003. Thus, the manager of a property with an accounting period ending on 31 December 2003 will be required to consult on such expenditure on agreements to be entered into from 1 January 2004.

In properties with unequal service charge contributions, the manager will have to calculate whether the proposed expenditure will cost any one of the service charge payers more than £100 for the year: if so, he is required to consult all service charge payers.

Stage 1 – Notice of Intention

Other than in public sector developments where public notice is required, the consultation procedure for qualifying long term agreements is set out in Schedule 1 to the Regulations. The first stage is for the landlord or manager to notify his intentions to enter into an agreement. The notice must comply with the following:

- Notice is to be given to each tenant, and to any recognised tenants' association (as defined by section 29, Landlord and Tenant Act 1985).
- The notice must describe the relevant works or services to be provided, or specify a place and times when the description can be inspected.

- If the description is made available for inspection, the place and time specified must be reasonable, and facilities (including copies) must be free of charge.
- The notice must include a statement of the reasons for the agreement.
- The notice must invite written observations, specify where observations should be sent, and state that observations must be made within 30 days from the date of the notice and the date when that period ends.
- The notice must also invite nominations of persons from whom the landlord should try to obtain alternative estimates (note: this is not to be an invitation for alternative estimates *per se*, presumably because of the risk of breaches of commercial confidentiality).

In the event that any observations are received during the allotted period, the landlord or manager must 'have regard' to them (paragraph 10 of the Regulations). By implication therefore, there is no duty to have regard to observations made outside the period. Neither the Act nor the Regulations give a definition for 'regard', which is the same term as that employed in the original version of section 20. The sanction for failing to treat this duty seriously will not be immediate, but it could be severe.

If, on a subsequent challenge to the expenditure at the LVT, the tribunal determines that the expenditure on the agreement was unreasonable and that it could have been more reasonable had proper regard been paid to observations, two things could happen: first, the tribunal is likely to reduce the recoverable expenditure to a more reasonable amount (which could happen in any event); second, the tribunal could take the view that the failure to have proper regard to observations amounted to non-compliance with the consultation procedure and thus further reduce the charges to the statutory limitation of a maximum of £100 for any tenant. By the time the LVT reached its decision, the manager could have committed himself to substantial outlay on the agreement which would now have to come from his own resources. In the case of an RMC or an RTM company, with no resources other than service charge income, this could represent a very serious setback.

Another point which arises generally in relation to leaseholders' nominations is that nominees may not meet the normal criteria which some landlords set for contractors, whether as policy or because they are required to do so (especially with public sector and social landlords). These criteria can cover such matters as public liability insurance, health and safety procedures, and equal opportunities policies. These are important matters, and it is likely that many landlords will seek to insist that their criteria are covered by any estimates they elect to put forward. Whether such an approach will be regarded as reasonable by the LVT will become clear as the cases develop.

For the sake of consistency, landlords in such a position may be prudent to notify both leaseholders and their nominated contractors of their criteria at the outset of any consultation procedures.

Stage 2 – Estimates

The rules for dealing with estimates from nominated contractors are set out in paragraph 11 of the Regulations, as follows:

- If a single nomination is received from a recognised tenants' association ('RTA') (whether or not any are received from individual tenants), the manager must seek an estimate from the nominee.
- If a single nomination is made by only one tenant (whether or not one is made by an RTA), the manager must seek an estimate from that nominee.
- If single nominations are made by more than one tenant (whether or not any are made by an RTA), the manager must seek an estimate from the person with most nominations, or, if there is no clear leader but there are two or more who tie for first place, from one of those. If the result is not even that clear (for example there could five nominees with one vote each), an estimate must be obtained from one of them. The choice then obviously rests with the manager.
- Finally, if multiple nominations are received from any tenant and more than one from the RTA, the manager must request an estimate from at least one person nominated by a tenant and at least one nominated by the RTA.

This does not quite seem to answer all possible eventualities: for instance, there appears to be no indication of how to deal with a single nomination from a tenant and multiple nominations from an RTA. It must be hoped that common sense will prevail in such circumstances.

Stage 3 – The landlord's proposals

As will have been gathered by now, the new section 20 procedure will be longer drawn-out than before. The landlord or manager has to wait 30 days from his initial notice of intention. Then, if any nominations are received during the 30-day period, there will be further time spent in obtaining estimates from nominated contractors: this could take several weeks in itself. Following that (or in any event, if no nominations were received), the landlord or manager must publish his proposals pursuant to paragraph 12 of the Regulations.

This part of the process requires the landlord or manager to prepare at least two proposals in respect of the goods or services to be supplied. At least one of them must relate to a person wholly unconnected with the landlord. Further, where an estimate has been obtained from a tenant's nominee, that estimate must be the basis of one of the proposals.

Each proposal must contain:

- A statement of the works, goods or services to be provided.
- The name and address of each party to the proposed agreement (except the landlord) and a statement of any party's connection to the landlord.
- The definitions of 'connection' are set out in paragraphs 2(1) and 12(6) of the Regulations, and are examined briefly below.
- Where practicable, an estimate of each tenant's contribution to the cost of the proposed agreement.
- Where it is not practicable to do so, but it is reasonably practicable to estimate the total expenditure, an estimate of that total.
- Where neither of the last two estimates is practicable, but it is reasonably practicable to estimate the current unit cost or hourly or daily rate, a statement of those figures.
- If any of the proposed works or services involve the appointment of an agent to undertake any of the landlord's management obligations under the leases (in other words, a managing agent), a statement that the proposed contractor is or is not a member of a professional body or trade association (such as ARMA or ARHM) and subscribes or does not subscribe to any relevant code of practice or voluntary accreditation scheme. If the contractor is a member of a professional body or trade association, the name of the organisation must be given.

- A statement of any provisions to vary or determine the charges under the proposed agreement.
- A statement of the intended duration of the agreement.
- Where observations were received to which the landlord or manager had to have regard (under paragraph 3 of the Regulations), a summary of the observations and his response to them.

When it comes to ascertaining whether a proposed contractor is connected to the landlord (see above), it is first necessary to establish whether the parties are companies, partnerships or individuals. With trading names, this is not always straightforward, but it should be in the knowledge of the landlord or manager. The thrust of the rules is that a connection is to be assumed if any of the individuals concerned is a director, manager or partner in the business of the other party, or is a close relative of such a person. A 'close relative' is a spouse or cohabitee, parent, parent-in-law, son or son-in-law, daughter or daughter-in-law, brother or brother-in-law, sister or sister-in-law, step-parent, step-son or step-daughter.

Notice of the proposals must be given to each tenant and any RTA. The notice is to be accompanied by a copy of each of the proposals, or must specify a place and times at which they may be inspected. The facilities for inspection are to be the same as set out above in respect of the notice of intention.

The notice must also include an invitation for written observations, specifying where they should be sent, the period allowed (another 30 days) and the date when the period for observations will end. Once again, regard must be had to observations received within the 30-day period.

Stage 4 – the landlord's decision

When the landlord or manager has finally decided which of the proposed agreements to take up, unless the contractor is to be a tenant's nominee or the one who submitted the lowest estimate, he has 21 days within which to give written notice to each tenant and any RTA stating his reasons for his election or specifying a place and times where a statement of his reasons may be inspected. In addition, when observations upon his proposals had been received within the 30-day period, the notice must include a summary of the observations and his responses to them (or details of how the observations and responses can be inspected). Any facilities for inspection are to be given in the same way as regards the notice of intention.

In summary then, the manager of leasehold premises has to consult for 30 days on his intention to enter into a qualifying long term agreement, seek estimates from nominees thereafter, and then consult for a further 30 days on at least two specific estimates. This process will take a very minimum of 60 days, but is more likely to last for three months or more. A great deal of work will be required with associated management, stationery and postal costs. Management fees will almost certainly be increased to allow for this. Obviously, the process (and costs) will have to be repeated for every relevant type of agreement.

The burden on the likes of RTM companies and RMC's will be severe, but this is unlikely to prove a reasonable ground for dispensation from the consultation requirements by the LVT under section 20ZA(1).

The consultation requirements where public notice is required

Public notices are needed in relation to certain public sector tenancies pursuant to the Public Works Contracts Regulations 1991 (SI 1991/2680), the Public Services Contracts Regulations 1993 (SI 1993/3228) and the Public Supply Contracts Regulations 1995 (SI 1995/201). The consultation requirements in such cases are set out in Schedule 2 to the Regulations.

Managers who are subject to these regulations are advised to refer to the relevant guidance notes published by (among others) the Leasehold Advisory Service (LEASE) and the Chartered Institute of Housing.

Qualifying works

Grasping the concept of consultation on qualifying works will be an easier task for those who have been familiar with the original version of section 20. The works in question are 'works on a building or any other premises' (section 20ZA(2)), and they qualify for consultation if the cost results in a contribution by any tenant of more than £250 (section 20(3) with the figure being inserted by paragraph 6 of the Regulations).

As with qualifying long term agreements, the landlord or manager will have to calculate whether the contribution of any single tenant is to exceed the £250 threshold, and then carry out consultation with all the service charge payers.

The consultation requirements for qualifying works are set out in paragraph 7 and Schedules 3 and 4 of the Regulations. Schedule 3 deals specifically with qualifying works which arise under a qualifying long term agreement.

The works affected by the transitional provisions of Regulation 7(3)(a) are:

- Any works which are carried out (by which Regulation 7(3)(a) presumably refers to works which commence) two months after the commencement date for the Regulations (31 October 2003), even though the agreement was entered into before the Regulations came into force.
- Works under an agreement for over 12 months where public notice was given before the Regulations came into force which are carried out (presumably commenced) after the commencement date of 31 October 2003.

Paragraph 3(2) of The Commonhold and Leasehold Reform Act 2002 (Commencement No. 2 and Savings) Order 2003 sought to clarify the position regarding an agreement for specific works which had already been the subject of consultation under section 20 as it stood before 31 October 2003, but which had not commenced before that date. It is clear that such works which had commenced by that date would be unaffected by the new procedures. It would appear that works which had not started before 31 December 2003 should be subject to a further round of consultation under Schedule 3 of the Regulations (as with qualifying works under qualifying long term agreements – see below). The lack of clarity is caused by the use of the phrase 'carried out' rather than 'commenced' in this context.

Unless and until this point is clarified by the government or through case law, it may be prudent either to go through the new section 20 procedures or to apply to the LVT for dispensation. It is unlikely that it was Parliament's intention to duplicate consultation, but the criteria have changed considerably, and the risk of being held subsequently to have failed to comply with the consultation requirements is so significant that the point should be considered by managers finding themselves in this position.

Consultation requirements for qualifying works under qualifying long term agreements

Written notice of the landlord's or manager's intention to carry out qualifying works must be given to each tenant and any recognised tenants' association ('RTA'). The notice must:

- Describe in general terms the proposed works or specify a place and times whereby the description may be inspected.
- State the reasons for the works.
- Estimate the total expenditure to be incurred by the landlord or manager in connection with the works (which would appear to include management costs, surveyors' fees and other incidental costs as well as the anticipated contract price).
- Invite written observations upon the proposed works or the estimated price.
- Specify the address where observations should be sent, the period allowed (30 days from the notice), and the end date for observations.

As with previously referred to notices under section 20, inspection facilities are to be free of charge and the place and times for inspection must be reasonable.

The landlord or manager must have regard to any observations received from tenants or RTAs during the 30-day consultation period. He must then respond in writing to such observations within 21 days of their receipt.

Subject of course to any applications to the LVT, the landlord or manager may then proceed with the works; there is no need for further consultation as the Act presumes that full consultation would have taken place already on the overall long term agreement concerned.

Consultation requirements for qualifying works (under public notice)

The provisions relating to consultation over works affecting public sector tenancies in the circumstances where a public notice is required are set out in Schedule 4 Part 1 of the Regulations.

Consultation requirements for qualifying works (without public notice)

Schedule 4 Part 2 relates to all private sector properties and to the public sector when no public notice is required for the qualifying works.

Stage 1 – Notice of Intention

The landlord or manager must give written notice of the intention to carry out qualifying works (ie where the contribution of any tenant will exceed £250) to each tenant and any RTA. The notice must:

- Describe the works in general terms or specify a place and times when the description may be inspected.
- State the reasons for the works.
- Invite written observations regarding the works and specify where they should be sent and the date when the consultation period will end (30 days from the notice).
- Invite each tenant and any RTA to nominate (within the 30-day period) a person from whom the landlord or manager should seek an estimate for the works.

Facilities for inspection must be provided on the same basis as the other notices referred to above.

The landlord or manager must have regard to observations received within the consultation period.

Stage 2 – Estimates

Regulation 38 sets out the procedure regarding estimates in a similar but not identical way to that for qualifying long term agreements. The principal points are as follows:

- If a single nomination is received from a recognised tenants' association ('RTA') (whether or not any are received from individual tenants), the manager must seek an estimate from the nominee.
- If a single nomination is made by only one tenant (whether or not one is made by an RTA), the manager must seek an estimate from that nominee.
- If single nominations are made by more than one tenant (whether or not any are made by an RTA), the manager must seek an estimate from the person with most nominations, or, if there is no clear leader but there are two or more who tie for first place, from one of those. If the result is not even that clear (for example there could five nominees with one vote each), an estimate must be obtained from one of them. The choice then obviously rests with the manager.
- If multiple nominations are received from any tenant and more than one from the RTA, the manager must request an estimate

from at least one person nominated by a tenant and at least one nominated by the RTA.

- Having obtained the estimates from any nominees (and, presumably, his own proposed contractors) the landlord or manager is required to supply (free of charge) to each tenant and any RTA a statement (referred to in the Regulations as the 'paragraph b statement') setting out the estimated cost from at least two of the estimates (making all of them available for inspection) and a summary of the observations received and his responses to them.
- At least one of the estimates shown in the statement must be from a person wholly unconnected with the landlord. ('Connected' is defined in the same terms as referred to above in the context of the landlord's proposals for qualifying long term agreements.)
- If any estimates were received from tenants' nominees, those estimates must be among those included in the statement.

Stage 3 – Inviting further observations

With the statement setting out the estimates, the landlord or manager must give a notice with details of where and when the estimates may be inspected (free of charge), and invite each tenant and any RTA to make written observations on the estimates. The invitation must specify an address where observations should be sent and state the period allowed for this purpose (30 days from the notice) and the end date of that period.

The landlord or manager shall have regard to observations made within the 30-day consultation period.

Stage 4 – the landlord's decision

Unless the chosen contractor is a tenant's nominee or is the person who submitted the lowest estimate, the landlord or manager has to give notice to each tenant and any RTA stating his reasons for his selection of contractor or say where and when a statement of reasons may be inspected. Furthermore, the notice must contain a summary of the observations received during the 30 day consultation period and his responses to them.

Summary

Once again, it is clear that the process for consulting on qualifying works will take 60 days at the very least, and will involve a good deal of time and expense. It is conceivable that a manager of leasehold premises may have to carry out several consultation exercises on long term agreements and qualifying works during a single year. Because the qualification criteria are triggered when any one tenant's contribution exceeds the threshold (£100 for long term agreements or £250 for building works), there will be additional expense of management time in calculating whether consultation applies, especially in a development where the service charge contributions vary.

The administrative burden upon RTE companies, RMCs and RTM companies will be substantial, and should form part of their budgeting processes. When leaseholders are considering exercising their rights such as the Right to Manage, and planning their funding and management arrangements, the increased level of consultation is another factor they and their advisors should be taking into account.

A practical problem for RTM companies, which has yet to be resolved, is the need to have contracts in place immediately from the acquisition date, which will need to be entered into before the RTM company is of the class of managers required to carry out consultation. It is perhaps likely that Leasehold Valuation Tribunals will look kindly on applications for dispensation from consultation in such circumstances.

Checklist for managers

The specific procedures for consultation on the various types of qualifying works, goods and services are detailed in the body of this chapter, but there are a few points which apply generally.

1. Ensure property managers are fully trained on the relevant criteria and procedures. Omissions will prove expensive.
2. Ensure property managers have access to a fully up-to-date copy of section 20 of the 1985 Act, as amended.
3. Have a calculation readily available to compute the service charge contributions and the overall expenditure on particular items which will trigger the consultation criteria.
4. Check which contracts and services at your property will be affected.

5. Check which contracts planned before 31 October 2003 will be caught by the transitional provisions.
6. If you are a superior landlord or you are acting for one, note that his agreements will be subject to the consultation requirements, and adjust your procedures accordingly.
7. If an application to the LVT for dispensation seems appropriate, lodge it as early as possible.
8. Note that consultation will now take longer than before, and will cost more in terms of management time and administration, and budget accordingly.
9. Check whether any contractors have any connection with the landlord or manager in the context of the definition now available.
10. Generally, ensure systems reflect the various duties throughout the consultation procedures and fit in with the new time constraints.

Chapter 12

The New Accounting Regime

Traditionally, there have been almost as many different accounting practices operated by property managers as there are leasehold properties. Some (but not many) are driven by the contents of the leases, some use the various codes of practice as a guide, and others are dictated by the available computer packages. Some provide very little information voluntarily; others overwhelm leaseholders with statistics and tables.

By the provisions of a number of sections in the 2002 Act, Parliament has sought to introduce a far more uniform accounting procedure for service charges. The new requirements had generally not been implemented by the end of 2003 however, because of the immense practical difficulties which need to be overcome. The majority of this chapter therefore deals with the principles laid down by the Act, which still need to be fleshed out by secondary legislation.

Statements of account

Section 152 of the 2002 Act replaces section 21 of the Landlord and Tenant Act 1985 (the 1985 Act). Under the old section 21, leaseholders were empowered to request various details behind service charge accounts once they had been served. The new version turns that around to put the onus on the landlord (the definition of which includes managers, resident management companies, RTM and RTE companies) to provide prescribed detailed information up-front in statements of account.

The new section 21 (as substituted by section 152 of the 2002 Act) provides that an annual service charge statement must deal with the following:

- Service charges of the leaseholder and the leaseholders of 'dwellings associated with his dwelling' (which is intended to mean those which contribute to the same service charge fund).
- Relevant costs relating to those service charges (see Chapter 10 for a definition of 'relevant costs'; essentially it is all those expenditure items which are attributable to the service charge).

- The aggregate amount standing to the credit of the leaseholders concerned at the beginning of the accounting period (a maximum of 12 months) and at its end.
- 'Related matters'.

(No doubt the phrase 'related matters' will be amplified by the awaited secondary legislation.)

The account statement must be given to each leaseholder within six months of the end of the accounting period (section 21(2)). Superficially, this does not seem to be a particularly onerous time-limit; however, there will be occasions when not all the information will be readily ascertainable (for example, contractors' final invoices). As we shall see from the new section 21A (below), there is a severe penalty for late delivery of the account, so attention must be paid to its prompt production.

Further, the account must be certified by a qualified accountant to the effect that, in his opinion, it deals fairly with its subject-matter and is sufficiently supported by accounts, receipts and other documents which have been produced to him (section 21(3)(a)). Individual leases will frequently set out contractual requirements for certification of accounts, often by managing agents. This new subsection should be seen as a supplement to the lease's requirements, not a replacement. Accordingly, if the Act requires an accountant's certificate, and the lease calls for certification by the managing agent, it would be prudent to supply both.

Additionally, each statement of account must be accompanied by a summary of leaseholders' rights and obligations in relation to service charges (section 21(3)(b)). This summary will be prescribed by secondary legislation in due course. The requirement to produce it is strict, so managers should keep a watchful eye for this provision coming into force, and be ready to up-date the formats of their account statements.

Similarly, the regulations yet to be implemented may provide prescribed forms or contents for the accountants' certificates and the statements of account themselves.

With regard to service of the account, if the leaseholder has notified the landlord or his agent of an address in England or Wales (other than the flat) for service of such documents, that address must be used (section 21(6) and (7)). It may be unlikely that a leaseholder has specified an address simply for service charge accounts. Both from a practical point of view and to avoid unnecessary arguments later, managers will be well-advised to use alternative addresses notified to them for service of any documents relating to service charge issues, or indeed any matters which they need to communicate to the leaseholder (unless legislation expressly requires service at the flat).

The right to withhold service charge payments

Section 152 of the 2002 Act also introduces a new section 21A into the 1985 Act. This new section affords a new statutory right for leaseholders to withhold payment of service charge contributions in certain circumstances, specifically if:

- The landlord has failed to supply any document within the time-limit stipulated in the body of section 21.
- The form or content of any document supplied under section 21 does not conform 'exactly or substantially' with the requirements to be imposed under the secondary legislation.

There are limitations on the exercise of this right however:

- The maximum figure which may be withheld is the total of the amount paid by the leaseholder during the current accounting period and the amount standing to his credit at the beginning and the end of the accounting period (which may be a difficult sum to calculate, especially if the account statement has yet to be produced).
- If payment has been withheld because of late delivery of a document, the right ceases when the document has been supplied (albeit late, but subject to other legislation on limitation, including the '18-month rule' in section 20B of the 1985 Act).
- In the event of documentation which did not comply with the regulations, the right ceases when replacement documentation which does comply has been delivered.
- The right also expires if the Leasehold Valuation Tribunal has determined that the landlord had a reasonable excuse for non-compliance.

While a leaseholder is lawfully exercising the right to withhold payment, any enforcement or penalty provisions in his lease will be of no effect (section 21A(5)).

Service charge demands

Section 153 of the 2002 Act introduces a new section 21B into the

1985 Act. This section requires that any service charge demands (which would include requests for interim payments or monies on account) must be accompanied by a summary of leaseholders' rights and obligations relating to service charges. The form and content of the summary will be prescribed by secondary legislation.

Failure by the landlord to comply with this requirement will give the leaseholder a further right to withhold payment, and no enforcement or penalty provisions in the lease will be effective while the right subsists. Unlike section 21A, the new section 21B does not state that the right to withhold payment lapses when the landlord's non-compliance has been rectified, but that can be inferred because a properly submitted demand will not give rise to the right. The landlord may have a problem in enforcing payment however if a correctly constituted demand is not served until after the due date for payment in the lease has passed. In such a case it may be necessary to refer the matter to the LVT.

Inspection of documents

The legislation governing leaseholders' rights to inspect documentation supporting service charge accounts is contained in section 22 of the 1985 Act. Section 154 of the 2002 Act substitutes section 22 by a complete replacement, although it is similar to the original.

By the new section 22, a leaseholder may give written notice to the landlord requiring reasonable facilities for inspecting or copying accounts, receipts or other documents relating to the matters set out in a statement of account given under section 21. Alternatively, the leaseholder may request the landlord to take copies and send them to him or provide reasonable facilities for inspection. Although it is not spelt out in the section, it is probable that the LVT will have the power to determine what is 'reasonable'.

As before, a recognised tenants' association (as defined under section 29 of the 1985 Act) may represent the leaseholder in obtaining the requisite facilities and information.

A notice under section 22 must be served within six months of the date by which the leaseholder should have received the relevant account statement under section 21 (section 22(3)); in other words, it must be served within 12 months of the end of the accounting period. If the notice is served later, presumably the landlord can ignore it. If the account statement is served late however, the six-month period for service of the section 22 notice does not start to run until the account is served.

The notice may be served upon the landlord, the person who receives the rent on his behalf, or 'an agent of the landlord named as such in the rent book or similar document' (section 22 (5)). Rent books are rarely employed in long leases. It is not made clear what might constitute a similar document to a rent book. Presumably the definition is intended to cover managing agents, but whether such a construction will be placed upon it by the courts or the LVT remains to be seen. Any recipient of a notice who fits the description is required to pass it on to the landlord 'as soon as may be'.

The landlord must comply with a section 22 notice within 21 days of receipt. Facilities for inspection must be supplied free of charge to the individual making the request, but consequential costs may be claimed as management costs (section 22(7)), always provided of course that the lease allows for the recovery of management costs. The landlord may charge for anything other than facilities for inspection (section 22(8)), which must mean that he can make a reasonable charge for copying and postage under section 22(1)(b).

Service charge information held by superior landlords

In the event that a landlord requires information from a superior landlord in order to complete the statement of account under section 21, or to provide documents following a section 22 notice, he may invoke the participation of his superior landlord. This could arise, for example, where a residential development is a subsidiary part of a larger scheme, perhaps including retail or commercial elements. In such a case, it is likely that the landlord of the residential area is the tenant to a head-lease, under which he is liable to contribute to the estate expenditure (a contribution which is then divided up among the residential leaseholders).

Schedule 10 to the 2002 Act introduces a substitute section 23 into the Landlord and Tenant Act 1985, which provides:

- A landlord preparing a statement of account under section 21 may give notice to his superior landlord requiring the relevant information, and the superior landlord must comply 'within a reasonable time' (section 23(1)(b)).
- Where a leaseholder has requested documents under section 22, and those documents are held by a superior landlord, the immediate landlord who received the request must tell the

leaseholder that that is the case and supply the superior landlord's name and address. The leaseholder will then be able to apply section 22 directly upon the superior landlord (section 23(2)).

Commercial landlords and their agents will have to become accustomed to responding directly to requests for information from residential leaseholders, notwithstanding the absence of any direct contractual relationship. Failure to comply with these provisions will be an offence, punishable by a fine. Running the risk of prosecution over such administrative matters will be a new departure for commercial landlords.

Changes of landlord or leaseholder

Schedule 10 also introduces a new section 23A into the 1985 Act, which becomes relevant if a landlord or superior landlord disposes of his interest in the property before all his possible duties under sections 21 to 23 are completed. By section 23A(2) he remains responsible for those duties in so far as he is still in any position to do so. If, however, the new landlord is capable of performing the duties, his responsibility takes precedence (section 23A(4)).

Meanwhile, if a lease changes hands by assignment, a landlord's duties under sections 21 to 23A are unaffected by the assignment, save that a landlord is not required to comply with 'more than a reasonable number of requirements imposed by any one person' (section 24 of the 1985 Act as amended by Schedule 10 to the 2002 Act). In other words, a landlord has duties under these sections in relation to both outgoing and incoming leaseholders, but only to the extent that they are reasonable. In practice, this is only likely to apply in and around a conveyancing transaction which takes place between accounting periods.

Service charge disputes

Section 155 of the 2002 Act came into force on 30 September 2003, and expanded the jurisdiction of the LVT in the service charge field through the insertion of a new section 27A into the 1985 Act. The effect of section 27A is to permit the LVT to go beyond determining simply what is reasonable in the service charge context; the tribunal may now establish what is payable. Accordingly, the tribunal will have to consider what is validly due under the lease and under statute, and it will no longer be necessary to select whether the court or the LVT is the most appropriate forum for resolving a service charge issue; neither will cases need to be switched between the two during the course of proceedings.

A fuller analysis of the LVT's new jurisdictions on service charges and other questions will be found in Chapter 13.

Designated service charge accounts

The amendments to section 42 of the Landlord and Tenant Act 1987 (the 1987 Act) as introduced by section 156 of the 2002 Act have been among the most controversial changes in the management area, and they have proved to be among the most problematic to implement. This is possibly because of the need to involve the banks and financial institutions in providing account systems which are structured in such a way that the requirements of the Act can be met in practice.

The fundamental principle enshrined in section 42 of the 1987 Act is that service charge contributions are held on trust by the landlord for the contributing leaseholders as beneficiaries. The landlord is therefore a trustee with fiduciary duties to leaseholders, which can be quite a burden, especially for small landlords and resident management companies (and now RTM and RTE companies).

That fundamental principle is unchanged by the amendments in the 2002 Act; rather it is strengthened and supplemented.

The amendments are carried through in section 156 by the insertion of new sections 42A and 42B into the 1987 Act. Section 42A provides that any credit balance in the trust fund must be held in a designated account at a relevant financial institution (to be more clearly defined in the secondary legislation). An account qualifies if:

- The institution has been notified in writing of the purpose of the account.
- No other funds are held in the account.
- It fits the description to be specified in the regulations.

Whereas it is normal for trust funds to be separately traceable, the innovation here for landlords and property managers is that the banking account concerned must be free of other funds. Many landlords and managing agents have operated client accounts similar to those held by solicitors, relying on the accuracy of their record-keeping to trace the funds of individual properties as and when required. The economies of scale flowing from this approach are obvious, and could be substantial. Such a system does not allow for the degree of transparency now required under the 2002 Act however. Significantly, during the consultation period leading up to the 2002 Act, the government recognised that there would be increased management costs from these measures, which were likely to be passed on to leaseholders.

The need for separate banking can be seen from the new section 42A(3), by which a leaseholder who contributes to the service charge fund may give written notice to the landlord (of whichever description) requiring facilities for inspecting or copying documents from which he can ascertain compliance with the requirement for a designated account, or requesting that the landlord sends copies to him (or arranges for his collection of copy documents). The other mechanical elements relating to inspection are the same as for documents under section 22 of the 1985 Act (as amended – see above).

If any of the contributing leaseholders have reasonable grounds for believing that there has been non-compliance, they may withhold payment of service charges, and no provision in their leases for enforcement or penalty will have effect while this remains the case. This is not the first time that the 2002 Act gives leaseholders a statutory right to withhold payment of service charges, but it is the first and only occasion when the right arises because of a leaseholder's reasonable grounds for belief. Clearly the landlord will be able to apply to the LVT for a finding on whether or not there was true compliance with the accounting requirements, but it would be a more difficult task for the LVT to determine whether the leaseholders concerned had reasonable grounds for believing that there had been non-compliance. This would involve an objective test of a subjective belief which falls more commonly to a court with criminal jurisdiction.

The difference between this and the other circumstances in which payment might have been withheld is that the LVT's decision in other cases would be effectively retrospective (assuming the tribunal found that there had been compliance), but in a case under section 42A the tribunal could find that there had been compliance but that the leaseholder had had reasonable grounds for believing that there had not been compliance. Consequently, the leaseholder could have been right to withhold payment even though the landlord had acted in compliance with the section up until the time of the LVT's decision.

The distinction would be relevant only in the context of the landlord seeking to exercise enforcement powers or add penalties (such as interest) for late payment.

There are additional sanctions available against a landlord who has failed to comply with the accounting requirements of section 42A, and these are contained in the new section 42B which states that failure to comply, without reasonable excuse, is an offence punishable by a fine not exceeding level 4.

In the event that an offence has been committed by a company or other body corporate (such as an RTM company for example) with the 'consent or connivance of a director, manager, secretary or other similar officer', or due to neglect by such a person, that individual will also be liable to punishment for the offence.

Failure to comply with sections 42 or 42B will also provide a further ground for an application to appoint a manager under Part 2 of the 1987 Act (section 24(2) of the 1987 Act, as amended by Schedule 10 to the 2002 Act).

As can be seen, the various penalties for failing properly to comply with the new statutory requirements for holding service charge funds can be severe on landlords, managers and individual directors and officers. Any party responsible for looking after these funds will need to ensure that their accounting procedures are watertight.

Estate management schemes

Owners of freehold properties who are required to contribute to estate service charges (so long as they are under the Leasehold Reform Act 1967 or the Leasehold Reform, Housing and Urban Development Act 1993) are given new rights by section 159 of the 2002 Act. Section 159(2) provides that variable estate charges are payable only to the extent that they are reasonable. Where the charge is fixed or is ascertainable through a formula specified in the documentation for the scheme, the LVT will have the power to vary the documentation.

In short, the law and procedures relating to such procedures (which came into force on 30 September 2003) will now mirror very closely the provisions for service charges payable by leaseholders (see under 'Service charge disputes' above, and Chapter 13 for a discussion of the LVT's new jurisdictions).

Summary

The 2002 Act imposes a considerable additional burden on the accounting practices of landlords and property managers. Landlords with fewer resources (such as residents' management companies and RTM companies, who are landlords for service charge purposes) are likely to find the new requirements particularly cumbersome. The risks of non-compliance are severe, as are the associated penalties. All managers of leasehold property should undertake a thorough review of their procedures to ensure compliance.

Checklist

1. Ascertain precisely which statutory requirements are in force when undertaking or planning any accounting tasks.
2. Ensure account statements meet the new requirements under section 21.
3. Ensure the supply of summaries of leaseholders' rights and obligations with each account statement and demand.
4. Ensure that the correct addresses for service are being used and are regularly updated.
5. Be prepared to meet promptly any requests for information or documents.
6. Ensure that all service charge monies are treated fully in accordance with the accounting requirements of section 42.

Chapter 13

The Role of the Leasehold Valuation Tribunal

The Leasehold Valuation Tribunal has seen a vast increase in its jurisdictions over the last few years. Now under the aegis of the Residential Property Tribunal Service, it has long since surpassed the role suggested by its name, that of valuations in leasehold enfranchisement cases. Since the relevant sections of the 1996 Housing Act came into force in 1997, the LVT has had a major part to play in service charge disputes, although it only had the power to determine whether service charges were reasonable, plus the jurisdiction to appoint a Manager for leasehold property under Part 2 of the Landlord and Tenant Act 1987 ('the 1987 Act').

This restriction on its powers saw an unfortunate overlap with the role of the courts, who retained decision-making over validity of service charges, consultation procedures and the like. As a result, cases were sometimes transferred backwards and forwards between the court and the LVT, causing inevitable inconvenience, delay and additional costs to the parties.

The Commonhold and Leasehold Reform Act 2002 ('the 2002 Act') brings in a number of changes to the role of the LVT, all of which will extend the tribunal's jurisdiction. In doing so, the potential for confusion over the comparative areas presided over by the LVT and the courts should largely disappear.

The Residential Property Tribunal Service has published a very helpful pamphlet entitled *Leasehold Valuation Tribunals: Guidance on procedure* which briefly describes the LVT's current jurisdictions as well as explaining its procedures.

Variations of leases

Before 30 September 2003, the jurisdiction for granting variations of leases rested wholly with the courts. From this date, the jurisdiction in respect of long residential leases under the Landlord and Tenant Act 1987 was transferred to the LVT (section 163 of the 2002 Act). This represents the clearest indication of Parliament's intention to vest the LVT with the authority to make judicial decisions in matters of interpretation and construction of legal documents, with the tribunals taking the place of the courts rather than just supplementing them.

It is hoped by many practitioners that this step will lead to a more flexible and practical approach to leases, as opposed to the more narrowly legalistic approach of the judiciary.

Another welcome step is the extension of the grounds in section 35 of the 1987 Act upon which applications for variations may be made. The extension is introduced by section 162 of the 2002 Act, which clarifies the insurance ground and brings in an entirely new ground: namely whether the lease makes satisfactory provision for 'an amount to be payable (by way of interest or otherwise) in respect of a failure to pay the service charge by the due date' (section 35(3A) of the 1987 Act as inserted by section 162(4) of the 2002 Act). This provision is expected to be of particular assistance to residents' management companies and RTM companies.

Section 162 also permits further grounds to be added by regulations by the Secretary of State from time to time. No additional grounds were introduced however by the Commencement Order implementing section 162 (SI 2002 No 1912).

Applications for variations for should be accompanied by a draft, although the LVT may alter the draft or substitute the tribunal's own preferred wording.

Administration charges

As previously discussed in Chapter 10, the LVT has now acquired the jurisdiction to determine the reasonableness of administration charges, whether fixed or variable. Schedule 11 to the 2002 Act gives the tribunals specific powers to vary leases to preclude unreasonable charges.

Service charges

There is a plethora of amendments to previous service charge legislation to be found in the 2002 Act, but the general thrust is clear: the LVT's jurisdiction in service charge cases is now unfettered. It is likely that the only cases concerning residential service charges which will still be heard regularly by the courts will be straightforward debt actions. Even then, a defended claim arguably should be transferred to the LVT for a determination of the dispute.

The kernel of the LVT's service charge jurisdiction is now to be found in section 27A of the Landlord and Tenant Act 1985, as inserted by section 155 of the 2002 Act. Briefly, this provides that applications may be made to the LVT to determine whether service charges (actual or proposed) are payable, and thus valid and lawful under statute and the terms of the lease concerned. If the tribunal holds the charges to be payable, it may then decide:

- By whom and to whom it is payable.
- The amount to be paid.
- The date and the manner payable.

Accordingly, the LVT can make any relevant decision that would previously have been expected from the court.

Furthermore, another loophole is closed by section 27A(2) and (5) which clarify that the LVT's protection is available even to a leaseholder who has paid the charge concerned (thus overruling the controversial decision in *R (on the application of Daejan Properties Ltd)* v *London LVT* [2002] 43 EG 187.

There are still some limited restraints upon the LVT's jurisdiction however. Section 27A(4) provides that no application may be made if the matter concerned:

- Has been agreed or admitted by the leaseholder (but payment is not to be construed as an agreement or admission – section 27A(5)).
- Has been determined already by the Court or through a post-dispute arbitration agreement.
- Has been referred or is to be referred for arbitration through a post-dispute arbitration agreement.

A 'post-dispute arbitration agreement' is one willingly entered into by the parties concerned once the subject-matter of the dispute has arisen (section 38 of the 1985 Act as amended by section 155(2) of the 2002 Act). Any other form of arbitration agreement, such as an arbitration clause in a lease, will be regarded as void for these purposes and will not deprive the LVT of its jurisdiction.

Among other service charge issues which will now attract the interest of the LVT are:

- A landlord's application for a decision that he has a reasonable excuse for failing to comply with the new provisions affecting statements of account (section 21A(4) of the 1985 Act, as inserted by section 152 of the 2002 Act).
- By implication, the LVT would be able to determine whether a landlord had complied with the new accounting requirements

contained in section 42A of the 1987 Act (as inserted by section 156 of the 2002 Act) when hearing an application for a determination of what service charges are payable, in a case where the leaseholder had withheld payment.

Generally, the new rules regarding service charge accounting are analysed in Chapter 12.

Estate management schemes

By section 159 of the 2002 Act, the LVT is given the jurisdiction to determine what is or is not reasonably payable in respect of estate management schemes under the Leasehold Reform Act 1967 or the Leasehold Reform, Housing and Urban Development Act 1993 (see also Chapter 12). This is another new departure, in so far as persons liable to pay such charges are freehold owners of their properties rather than leaseholders.

The means by which the LVT is to determine estate charges is generally very similar to its process in service charge cases.

Right to Manage

Another entirely new jurisdiction for the LVT is created out of the introduction of the Right to Manage by the 2002 Act. It is not necessary to apply to the LVT to exercise the Right to Manage *per se*, as this is a statutory right which does not require proof of default. The LVT will have a role in some cases however in checking and enforcing compliance with the RTM provisions in the Act.

The LVT's functions in relation to RTM are analysed more fully in Part 1, but they may be summarised as follows:

- Determining the validity of a Claim Notice following service of a counter-notice (section 84(3)).
- Determining a claim when the landlord is not traceable (section 85).
- Deciding questions concerning costs (sections 88 and 89).
- Applications by landlords (or leaseholders) to determine issues of reasonableness or appoint Managers as against RTM companies.
- Determining the amount of accrued uncommitted service charge funds to be transferred to an RTM company at the acquisition date for RTM (section 94(3)).

- Determining objections to approvals during the currency of RTM (section 99).
- Determining applications from RTM companies that the general rule that RTM cannot be exercised at premises which were subject to RTM in the previous four years should be disapplied (paragraph 5(3), Schedule 6).

Consultation procedures

The LVT's new powers in relation to statutory consultation procedures over major works and long term agreements may be of special interest to RTM companies. The new consultation requirements are discussed at length in Chapter 11, but the key point for LVT jurisdiction is that it is now the LVT rather than the court which has the say in establishing whether or not the procedures were properly followed, or if it is reasonable to dispense with them in certain circumstances.

In the past, the court's dispensation has generally been called for in the event of an emergency. From the point of view of an RTM company there is a new problem. The RTM company is responsible for providing services immediately from its acquisition date. Prior to that date, the company will have had no standing to carry out consultation under section 20 of the 1985 Act (as amended by section 151 of the 2002 Act).

It is probable that the LVT will be asked to give early guidance on whether it is reasonable to dispense with the consultation requirements in such cases. No doubt its decisions will be informed partially by the degree of support for RTM in a particular development, so it may not be easy to establish a hard and fast rule.

Appointment of managers

As we have seen in previous chapters, applications may be made to the LVT for the appointment of a Manager on wider grounds than before (especially Chapter 12 under 'Designated service charge accounts').

There is a further significant change in section 160 of the 2002 Act (amending sections 22 to 24 of the 1987 Act), whereby the LVT may now appoint a Manager under Part 2 of the 1987 Act to assume the management functions of a third party. This extension would include residents' management companies who do not own freeholds, and would also appear to cover Right to Manage companies.

Section 161 of the 2002 Act meanwhile amends section 21 of the 1987 Act by restricting the exemption of resident landlords from applications for the appointment of a Manager. Applications may now be made if at least half the flats in the premises are held on long residential leases. (In this context, 'resident landlords' and 'long leases' are as defined in the 1987 Act.)

Sections 160 and 161 came into force on 26 July 2002.

Landlord's choice of insurer

Section 165 of the 2002 Act tidies up paragraph 8 of the Schedule to the Landlord and Tenant Act 1985. The amendment extends the jurisdiction (now of the LVT) to determine an application on a landlord's nominated insurer to a 'nominated or approved' insurer (see Chapter 14).

Determinations of breaches of covenant

There is one further major expansion of the LVT's jurisdiction to be considered, and that is in the context of forfeiture or proposed forfeiture of a lease.

Since the introduction of section 81 of the Housing Act 1996, no forfeiture of a residential lease for non-payment of service charges has been permitted unless the breach has been agreed or admitted by the leaseholder or determined by the court, the LVT or arbitration. By a somewhat convoluted path through sections 168 to 171, the 2002 Act effectively brings forfeiture for breaches of any covenants (except payment of rent) into line with section 81, and gives the power to determine such breaches to the LVT in most cases.

The new provisions relating to forfeiture are examined in more detail in Chapter 16, but the essential points with respect to the LVT's jurisdiction are as follows:

- In order to exercise forfeiture for any breach of covenant (other than non-payment of rent) a landlord must first serve a notice under section 146 of the Law of Property Act 1925 (a 'section 146 notice').
- With regard to non-payment of service charges (whether or not reserved as rent) a landlord may not proceed to forfeiture until the breach has been agreed, admitted or otherwise determined.
- From the implementation of sections 168 and 170 of the 2002 Act, no section 146 notice may be served regarding any breach

unless and until the breach has been determined to have occurred by the LVT (unless the LVT's jurisdiction is excluded).

- The LVT's jurisdiction will only be excluded if the leaseholder has admitted the breach or it has been determined by a court or through a post-dispute arbitration agreement (see under 'Service charges' above).
- No section 146 notice may be served, nor any forfeiture proceedings for non-payment of service charges commenced until 14 days after the breach has been finally determined.

Assuming that the LVT's jurisdiction will be excluded in only a small minority of cases, the tribunal will have an extremely substantial role to play in filtering future forfeiture cases. The effect of this in practice is likely to be that many landlords will be deterred from attempting forfeiture at all, which is undoubtedly one of Parliament's intentions.

The LVT's new procedures

In order to carry out all its many and varied new functions, the tribunals will have to adopt more robust and streamlined procedures than before. Some of the new jurisdictions involve matters which will be of some urgency to landlords, managers and leaseholders alike. In order to facilitate increased and accelerated turnover of cases, the 2002 Act and the regulations under it have made a number of significant changes to the LVT's powers and procedures. Clearly, the tribunals will also have to increase in numbers and resources to be able to cope with their additional workloads.

The procedural changes are brought in through section 174 which introduces Schedule 12 to the 2002 Act. Schedule 12 and its supporting regulations came into force on 30 September and 31 October 2003. These contain a range of new powers and sanctions which go a long way towards giving the LVT real 'teeth'.

Applications

General applications to the LVT must now state:

- The name and address of the applicant.
- The name and address of the respondent(s).
- The name and address of any landlord or tenant of the premises concerned.
- The address of the premises.

- A statement that the applicant believes that the facts stated in the application are true (a 'statement of truth' as used now in court proceedings).

There may be additional requirements which vary according to the nature of the application. Any of the requirements may be dispensed with however if the LVT considers the included particulars and documents are sufficient to proceed, and that no other party is likely to be prejudiced (regulation 3(8) of the Leasehold Valuation Tribunals (Procedure) Regulations 2003 ('the regulations').

In the case of an application to vary a lease, the applicant is required to give notice of it to any person who the applicant has reason to believe is likely to be affected by the proposed variation, and to the respondent. On receipt of the notice, the respondent is also required to give notice to potentially affected persons (regulation 4).

Generally, the LVT will serve copies of the application and accompanying documents on the respondents, and any other person known to and considered by the tribunal as likely to be significantly affected (regulation 5(2)(b)). Service may be by advertisement.

Where numerous applications are made covering matters which are substantially the same, the LVT may decide to deal with one of them as a representative application. The rules are somewhat esoteric and are set out in full in regulations 8 to 10.

Fees

The LVT should not proceed with a case unless its fee is paid, and it will be treated as withdrawn if the fee remains outstanding for a month, unless the tribunal considers it reasonable to wait (regulation 7).

Parties

Any person may request to be joined to LVT proceedings as either applicant or respondent, but the tribunal has discretion in determining such a request (regulation 6).

Information

The LVT may serve a notice on any party to proceedings requiring the provision of any information which the tribunal reasonably requires. The notice will specify a period (at least 14 days) for compliance. Non-compliance renders the party concerned liable to a fine not exceeding level 3 (Schedule 12, paragraph 4.)

Dismissal

By regulation 11, the tribunals are given a new and welcome power of summary dismissal of applications which are considered frivolous, vexatious or otherwise an abuse of the process. Cases may be dismissed (in whole or in part) on a respondent's application or on the tribunal's own motion.

Before dismissal however, the tribunal must give notice to the applicant stating that it is minded to do so and the grounds, and giving a date (no earlier than 21 days hence) by which the applicant may request a hearing on the issue. If a hearing is requested it must take place (regulation 11(4)).

Pre-trial review

The LVT may hold a pre-trial review on request or on its own initiative (regulation 12). The review will provide an opportunity for the tribunal to give directions for the 'just, expeditious and economical disposal of proceedings' (regulation 12(3)(a)). The regulations also provide implicitly that the review should be used to narrow the issues where possible by encouraging the parties to make any reasonable admissions or agreements, and recording them (regulation 12(3)(b) and (c)).

A pre-trial review may be conducted by a single tribunal member.

Determinations without a hearing

Alternatively, the LVT may now determine a case without a hearing (regulation 13), but only if one of the following circumstances applies:

- The respondent states in writing that he does not oppose the application.
- The respondent withdraws his opposition to the application.
- The parties agree in writing to such a determination.

If such a step is proposed, the LVT must notify the parties and invite written representations and set out the relevant procedure (regulation 13(2)). However, either party may request a hearing (which they must then have).

In the event that a matter proceeds to be determined without a hearing, it may be dealt with by a single tribunal member.

Hearings

The rules for full hearings are contained in regulation 14. They will be generally in public, although the tribunal may decide to hear a particular case in private, if the circumstances demand it. Each tribunal is empowered to establish the procedure for the conduct of a hearing, but regulation 14 (7)(b) lays down that a party may represent himself or appoint a representative (whether or not legally qualified). Each party will be entitled to give evidence and to cross-examine other parties' witnesses.

If a party does not appear, the tribunal may proceed in his absence if satisfied that proper notice of the hearing was given (regulation 14(8)).

Decisions may be given at the hearing or subsequently. Reasons should be given with written decisions, but are not required if an oral decision was handed down at the hearing (regulation 18).

Inspections of premises

At its discretion, the tribunal may inspect the premises which form the subject-matter of an application, or any comparable premises, on notice to the parties (regulation 17). Subject to any relevant consent being obtained, the parties may be invited to attend an inspection.

Costs

Although not on a par with the courts yet, the LVT's powers to award costs are advanced considerably by paragraph 10 of Schedule 12. The power arises only against a party whose application is dismissed as frivolous, vexatious or otherwise an abuse of the process, or if he has acted 'frivolously, vexatiously, abusively, disruptively or otherwise unreasonably' (paragraph 10(2)). It can be foreseen that the power will be used sparingly, but if it is, the maximum which the defaulting party can be ordered to pay to other parties is £500 (or such other sum as may be fixed from time to time).

Paragraph 10(4) of Schedule 12 provides that no-one can be required to pay another's costs incurred in LVT proceedings except by determination by the LVT within its limited powers. This would seem to suggest that contractual provisions for costs (such as a covenant in a lease covering the costs of enforcement or forfeiture) will be of no effect in so far as costs at the LVT are concerned. If this interpretation is correct, there could be a significant impact on landlords using enforcement through the LVT (as a preliminary to forfeiture for example), especially residents' management companies and the like who have no independent resources to fund enforcement actions. This situation will be examined in greater depth in Chapters 16 and 17.

Enforcement

A decision of the LVT may be enforced in the same way as orders of the court, subject to permission from the county court.

Appeals

Appeals from LVT decisions are strictly limited by section 175 of the 2002 Act. Appeals lie to the Lands Tribunal, but only with the permission of the LVT or, failing that, the Lands Tribunal. An appeal must be lodged within 21 days of the decision concerned (regulation 20). The LVT has the power to extend this and other time limits set out in the regulations, but parties must make their requests for extensions within the original time-limits (regulation 24).

On determining an appeal, the Lands Tribunal may exercise any of the LVT's powers, and its decision is enforceable in the same way as an LVT decision. The Lands Tribunal may only award costs in the same way as the LVT, and subject to the same limitations (see above under 'Costs').

Chapter 14

Insurance

The 2002 Act has sought to improve the position of leaseholders in relation to insurance issues in two areas: where leases require leaseholders to place insurance through the landlord's nominated insurer; and where leaseholders contribute to the landlord's insurance costs in the same way as other service charge expenditure.

The landlord's nominated insurer

Parliament has long perceived a problem afflicting particularly long leasehold houses (and some flats and maisonettes) where the leaseholder places the buildings insurance, but is required by a covenant to do so through the landlord's nominated or approved insurer. By the use of such a covenant, a landlord (who has little interest in the property otherwise except for ground rent) is thought to be able to exploit the leaseholder in two ways:

- He can influence the level of the premium (from which he will extract commission, directly or indirectly).
- It is easy for uninformed leaseholders to neglect to perform the covenant correctly, and thus lay themselves open to forfeiture. Since such an insurance provision is not strictly within the service charge framework, the landlord could proceed straight to forfeiture without first having to have the insurance cover determined as reasonable by the LVT.

Section 164 of the 2002 Act seeks to cure this ill (although it was not in force at the time of writing). The section applies specifically to long leasehold houses, and provides that the covenant to insure through the landlord's nominated or approved insurer is of no effect if:

- The house is insured through an authorised insurer (one who does not contravene the prohibition contained in section 19 of the Financial Services and Markets Act 2000).
- The policy covers the interests of both the landlord and the leaseholder.

- The policy covers all the risks required to be covered by the lease.
- The policy covers at least the sum assured as required by the lease.
- The leaseholder serves notice of the cover as required by section 164(3).

Under section 164(3) the leaseholder is to give a 'notice of cover' to the landlord within 14 days from the existing renewal date or, if it has not been renewed, the date the cover took effect. If the freehold changed hands, an incoming landlord may request a notice of cover within a month of his acquisition, and the leaseholder must supply it within 14 days of the request.

The notice of cover must be in a form which has yet to be prescribed, but it must contain:

- The name of the insurer.
- The covered risks.
- The amount and period covered.
- Such further information as has yet to be prescribed.

The notice may be sent by post and, if it is, it must go to the landlord's address last furnished to the leaseholder under section 48 of the 1987 Act or, failing that, to the last address furnished under section 47 of the 1987 Act. If the landlord has supplied a specific address in England or Wales for service of notices of cover however, it is that address which must be used (section 164(9)).

Section 165 of the 2002 Act follows up with a minor tidying-up amendment to paragraph 8 of the Schedule to the Landlord and Tenant Act 1985. The amendment extends the jurisdiction (now of the LVT) to determine an application on a landlord's nominated insurer to a 'nominated or approved' insurer.

Insurance as a service charge

It must be remembered that where leaseholders contribute to the landlord's costs of insurance, insurance will be a service charge item (as defined by section 18 of the Landlord and Tenant Act 1985 ('the 1985 Act')) however it is expressed in the lease, and however it is accounted for by the landlord. Many leases treat insurance as an entirely separate matter, for example insurance contributions are often reserved as an additional rent ('insurance rent'). None the less, for the purposes of the legislation affecting service charges, insurance will be dealt with largely identically to any other form of service charge expenditure.

Having said that, there are certain peculiarities to the way in which insurance is treated, and they relate principally to the way in which insurance documentation can be inspected. The Schedule to the 1985 Act deals specifically with such matters, and that Schedule is amended by paragraphs 8 to 13 of Schedule 10 to the 2002 Act. The amendments are generally routine and consequential, but one or two points are significant.

The request for a summary of the insurance is now to be given by written notice, and the landlord's response is required within 21 days of receiving the notice (rather than one month). Notwithstanding the references to a 'notice', there is no prescribed form for this purpose.

Paragraph 3 of the Schedule to the 1985 Act (which deals with requests to inspect policy documents) is completely substituted. By the new paragraph 3, a leaseholder may give written notice to the landlord requiring reasonable facilities for inspecting the policy or associated documents and for taking copies, or requiring the landlord to take copies and send them to him or provide reasonable facilities for collecting them. The notice may be served on the leaseholder's behalf by a recognised tenants' association, who may also inspect on his behalf.

The notice may be served upon the landlord's agent 'named as such in the rent book or similar document' ('similar document' is not defined) or the person who receives the rent, and that person must pass the notice to the landlord 'as soon as may be'. The landlord must comply with the notice within 21 days of receipt by him (not 21 days of receipt by his agent).

Any facilities for inspection must be supplied free of charge to the person serving the notice, but the costs can be covered through the service charge (always supposing that the lease authorises such management costs to be put to the service charge). Any other costs (such as copying presumably) may be charged to the person making the request.

'Associated documents' which may be inspected include accounts, receipts or other evidence of payment of the premium.

Other amendments concerning insurance effected by a superior landlord or change of landlord reflect the new provisions relating to service charge information (see Chapter 12).

PART 3

GROUND RENT AND FORFEITURE

Unlike most previous service charge legislation, the Commonhold and Leasehold Reform Act 2002 has included measures which go to the roots of the relationship between landlord and tenant in long leases. Forfeiture has been the subject of various attacks by Parliament over recent decades, especially in the service charge context (see particularly section 81 of the Housing Act 1996), but it is a very long time since the subject of ground rent fell under the legislative spotlight.

The changes to the law which are introduced by the 2002 Act are not fundamental, but they are significant, and they will require new approaches and procedures by landlords and property managers.

Chapter 15

Ground Rent: The New Regime

Most leases provide that forfeiture may occur on the grounds of non-payment of rent, whether the rent has been formally demanded or not. Further, the Common Law Procedure Act 1852 stated that a lease may be forfeited if rent is outstanding for over six months whether or not it has been demanded. Hardly any leases require that rent has to be requested; the dates and amounts to be paid are set out in the lease and landlords have tended to argue that leaseholders should take responsibility for paying on the due dates without being reminded.

As a result, ground rent has had a value to landlords in excess of its pure monetary worth. With ground rent not subject to the same scrutiny as service charges, forfeiture could be threatened without restraint for relatively small sums if payments were missed. Although the vast majority of leaseholders were able to obtain relief from forfeiture with no great difficulty, it was often an expensive and disagreeable experience. In the worst cases, leaseholders could lose their homes or their investments. Occasionally, after a leaseholder's mortgage lender had paid off the landlord and added the cost to the loan, the flat or house was repossessed by the lender because the leaseholder was unable to meet the increased mortgage repayments.

Of course, much of the history laid out above has already become a thing of the past. In many new developments and extended leases rents have been converted to peppercorns. In addition, a very high proportion of landlords are now companies owned by the leaseholders, who are extremely reluctant to use such sanctions against their neighbours and fellow shareholders. There is still a residual peril to a considerable number of leaseholders however, and Parliament has elected to use the 2002 Act to provide further protection.

Notification that rent is due

Section 166 of the 2002 Act (once it has been implemented) will effectively block the use of forfeiture when rent has been neither demanded nor paid by providing that the liability to pay rent under a long residential lease will not arise unless and until the landlord has given notice that the rent is due.

The notice is to be in a prescribed form (not in force by the beginning of 2004) and must specify:

- The amount due.
- The date it is due to be paid.
- The due date under the lease (if different).
- Other information as prescribed by the subsequent regulations (which may well include a summary of the leaseholder's rights and obligations, as with service charge demands).

In addition, landlords will still have to comply with the requirements of sections 47 and 48 of the Landlord and Tenant Act 1987 ('the 1987 Act') as to the furnishing of details of the landlord's name and address.

The due date

Notwithstanding the date reserved by the lease for the payment of rent, the leaseholder's liability will not accrue until the notice has been served, and the notice must give a date no earlier than 30 days and no later than 60 days after the notice has been given. In any case, the date cannot be before the due date stated in the lease (section 166(3)). None of the landlord's powers of enforcement or penalty can be deployed in respect of the due date in the lease, unless the notified date is the same (section 166(4)).

Landlords and their agents will need to review their procedures for demanding rent. It would seem to be prudent for the delivery of notices for rent to be timed so that they are received by leaseholders perhaps 59 days before the due date in the lease.

Service

The notice may be sent by post (section 166(5)). Postage is not mandatory so, by implication, the notice may also be delivered by other means (such as insertion through the letter-box of the dwelling). If it is sent by post, it must go to the demised premises unless the leaseholder has notified the landlord of an alternative address in England or Wales for service of rent notices; if he has, it is that address which must be used (section 166(6)).

The rules relating to service still leave possible loopholes open to the less scrupulous landlord. For example, the requirement to use an alternative address only applies to postal service; the notice for rent can still be delivered personally to the demised flat, even if the landlord knows perfectly well that the leaseholder does not reside there.

Also, the landlord's obligation to use an alternative address arises only if the leaseholder has specifically requested its use for service of notices under section 166, on a strict construction of section 166(6). Further, the alternative address, to be effective, must be notified to the landlord. The Act does not expressly provide in this context for service upon the landlord's agent. It may well be that these potential loopholes will be closed peremptorily by the courts.

Set against the conceivable ways in which landlords may take advantage of the service provisions are the legal and practical difficulties which the requirements pose for them. From a practical point of view, there will be a considerable increase in the administration needed to collect in ground rents, particularly if the prescribed forms turn out to be as lengthy as suggested during the consultation process. It will be rare for leases to allow landlords to charge their costs in demanding and collecting rents; meanwhile the sums involved could render the whole process uneconomic. For residents' management companies ('RMC's') owning their freeholds, the notification requirements are likely to add to their overheads which they cannot recover through the service charge. While issuing new leases at peppercorns may be more attractive, many RMC's rely on their rental income to pay for the company's running costs, and sometimes to service loans.

The issuing of demands for rent also causes landlords problems in a legal context in cases where they are contemplating forfeiture for other breaches of covenant. In order to balance the draconian nature of forfeiture, the courts have placed various hurdles in the path of landlords over the centuries. One of the steepest obstacles is the doctrine of waiver of forfeiture, whereby a landlord can waive his right to forfeit for a breach of covenant if he performs an act or omission which can be construed as acknowledging the continuation of the landlord and tenant relationship. The clearest act of waiver has always been the demand or acceptance of rent which has fallen due since the breach concerned took place.

Although the vast majority of landlords do send out notifications when rent is due, some have adopted the practice of holding them back when they are aware of other, more serious, breaches of covenant in order to avoid waiving their rights. Landlords have been able to do this confidently in the knowledge that rent would continue to accrue whether it was demanded or not, which will no longer be the case (section 166(1)). Henceforth, landlords will have to choose between risking waiver and forgoing rental income.

Definition of rent

For the purposes of section 166, the definition of 'rent' does not cover any other sums of money which may be reserved by a lease with such wording as 'further' or 'additional rent'. Accordingly, service charges, insurance contributions, or administration charges which are treated as rent in some leases will not be subject to the notification provisions of section 166 (section 166(7)). Such payments are subject to their own procedural regimes (see Chapters 10, 12 and 14).

Forfeiture for small amounts

Even when the leaseholder's liability to pay rent has arisen on the notification that it is due (however and wherever it has been served), there will be restrictions on the landlord's ability to take forfeiture proceedings imposed by section 167. This section prevents forfeiture for debts of less than a prescribed figure (probably to be set at £350) or where the debt has been outstanding for a short period (expected to be put at three years). These provisions will be examined more fully in Chapter 16.

The only significant limitation to debt recovery proceedings (as opposed to forfeiture) is the cost-effectiveness of such action to enforce relatively small debts. That is a consideration which will depend upon a number of factors, including the financial circumstances of the debtor and the degree to which the landlord (which may be an RMC) can minimise costs by acting without lawyers or other fee-charging debt collectors.

Checklist for landlords

1. Check from time to time on the implementation date for section 166.
2. Be ready to issue notices for rent in the prescribed form.
3. Ensure rent demands continue to show the information required by sections 47 and 48 of the 1987 Act.
4. Review procedures for service of rent demands.
5. Review procedures for recovery of unpaid rents.

Chapter 16

Forfeiture: Obstacles and Alternatives

Before examining the latest changes to the law relating to forfeiture, it is worthwhile to look at the context. For many years, Parliament has been looking for an alternative to forfeiture of residential leases. Thus far, no replacement has been found (although commonhold may be seen as a by-pass in the long term), and it is accepted that there must be an ultimate sanction for fundamental breaches of the covenants in a lease while leases still exist, so legislation has concentrated instead on modifying and ameliorating this unpopular ancient remedy.

The latest efforts at restricting the use of forfeiture are contained in the Commonhold and Leasehold Reform Act 2002 (although they are not expected to come into force until Spring 2004). The 2002 Act attacks forfeiture from a number of different angles. All forfeitures for monetary breaches are restrained where only small sums are outstanding for short periods. Most forfeitures will be subject to prior determinations that breaches have occurred. Forfeiture for unpaid service charges specifically become subject to additional restraints.

The law is changed further by the allocation of a principal role to the Leasehold Valuation Tribunal (LVT), whose involvement has been marginal in the past.

As we shall see, the cumulative effect of the changes will be to make forfeiture more difficult, expensive and long drawn-out, with the end result that the weapon will be blunted (which is presumably Parliament's intention).

Forfeiture for small amounts and short periods

By section 167 of the 2002 Act, no residential lease may be forfeited if the only breach of covenant alleged is the non-payment of rent, service charges or administration charges (or a combination of them), and:

- The amount in question does not exceed a prescribed sum

(expected to be £350, but in any event no more than £500 (section 167(2)); or

- The arrears are not restricted to sums which have been outstanding for less than a prescribed period (expected to be three years).

(The prescribed figure and period will be filled in by regulations, but the Government has been working on the basis of £350 and three years.)

The section is not particularly clearly drafted, but it appears that the Government's advice is that the expression 'or a combination of them' will not preclude a forfeiture action if the aggregate of the arrears exceeds the prescribed sum. Accordingly arrears of £100 rent, £100 administration charge and £200 service charge (amounting to £400 in total) will not be caught, so long as at least one item has been unpaid for more than the prescribed period. It remains to be seen how the courts will interpret the section however.

Another area which may require some elucidation is the way in which the two elements (amount and period) work with each other. The likeliest interpretation is that they go together, so that arrears of say £300 can only provide a ground for forfeiture if part of it has been outstanding for over three years, but there is no minimum period for arrears of say £600. Section 167(1) could conceivably be read however so that no forfeiture for less than £350 can ever take place, and no forfeiture for even a large amount can be launched unless part of the arrears is over three years old. The consequence of the latter interpretation may be unconscionable, but the courts' principal role is to construe what the legislation actually says before ascertaining what it is intended to mean. If it comes to fathoming Parliament's intention, it may help that section 167 is headed 'Failure to pay small amount for short period'.

The definition of service charge is as set out in section 18(1) of the Landlord and Tenant Act 1985 as amended; 'administration charge' is defined in Part 1 of Schedule 11 to the 2002 Act (see Chapter 10).

Finally, in computing the amount of the arrears for the purpose of establishing whether they are caught by section 167(1), any default or penalty charge added because of the fact of the arrears is to be disregarded (section 167(3)).

Prior determination of breach of covenant

Section 168 of the 2002 Act applies to cases where the forfeiture process has to be commenced by the service of a notice under section 146 of the Law of Property Act 1925 (a 'section 146 notice'), except in the circumstances of unpaid service charges or administration charges (which are dealt with in section 170 of the 2002 Act). Section 146 of the 1925 Act requires that a landlord intending to exercise a right of forfeiture must give notice to the leaseholder in the case of any breach of covenant other than payment of rent (or monies reserved as rent by the lease, as service charges sometimes are).

The purpose of a section 146 notice is to give the leaseholder an opportunity to apply for relief against forfeiture on compensating the landlord for the breach, which includes paying the landlord's costs. Most leases contain a covenant by the leaseholder to pay costs in such circumstances, sometimes on a full indemnity basis. It is perceived that some landlords have abused the section 146 procedure by insisting on excessive compensation and costs.

By section 168 of the 2002 Act, no section 146 notice may be served in respect of any breach of covenant unless certain criteria are satisfied:

- The LVT has finally determined on an application by the landlord that a breach has occurred.
- A court or arbitration (under a post-dispute arbitration agreement) has finally determined that a breach has occurred.
- The leaseholder has admitted the breach.
- In the event that a determination was necessary, fourteen days have passed since the determination and the breach has not been remedied within that period.

No application may be made to the LVT if the matter in question has already been determined or it has been, or is to be, referred to arbitration under a post-dispute arbitration agreement. A 'post-dispute arbitration agreement' is one which both parties have entered into willingly knowing that a dispute exists. Any other arbitration agreement (such as an arbitration clause in a lease) will be void for this purpose (section 169(1)).

A final determination by a court or LVT means not only that a decision has been made, but that any appeal has been dealt with or abandoned, or that the time for bringing an appeal has expired (section 169(2) and (3)). Appeal time-limits vary between the court and the LVT, but a typical period would be 21 days, so that a leaseholder who does not appeal effectively has around five weeks to remedy his breach before a section 146 notice can be served by the landlord.

Because section 146 of the 1925 Act does not apply to rent, and section 168 of the 2002 Act does not apply to service charges or administration charges, the breaches of covenant covered by section 168 are of a limited nature, but they could still be very significant matters. Covenants affected would include, for example:

- Repairs.
- Unauthorised structural alterations.
- Nuisance and annoyance to neighbours.
- Insurance provisions, such as acts which might render insurance cover void.
- Sub-letting.
- Unauthorised uses of the premises.
- Performance of easements, such as interference with rights of way.

The delay in dealing with such breaches of covenant occasioned by the need to have their existence determined by the LVT before even a section 146 notice could be served (let alone the commencement of forfeiture proceedings) could have serious consequences upon landlords (including, for example, residents' management companies) and other leaseholders.

Another by-product of section 168 will be the effect on costs. This effect will be felt particularly keenly by residents' management companies (RMCs), although all landlords will be subject to it, and by leaseholders in due course (whether or not they are the leaseholders in breach of covenant).

Given the extremely limited powers of the LVT when it comes to awarding costs (see Chapter 13), landlords must first look to recover the expense of proceedings under section 168 from the defaulting leaseholder in the context of his covenant to pay such costs. The ability to recover under the costs' covenant is restricted in two ways:

- The landlord can only collect from the leaseholder in accordance with the precise terms of the covenant. A typical costs covenant will allow for the landlord to charge for costs incurred 'of and incidental to the preparation and service of a notice' under section 146 of the 1925 Act. As no section 146 notice can be served until after a determination under section 168 has already taken place, the landlord's ability to recover costs up to that time does not start to run. Even more generous costs provisions in existing leases are unlikely to anticipate the circumstances of proceedings under section 168. Meanwhile, in some leases there are no reliable costs clauses at all.

- In any event, paragraph 10(4) of Schedule 12 to the 2002 Act states that no costs incurred before the LVT can be secured other than by a decision of the LVT (see Chapter 13). Arguably, this paragraph excludes the recovery of the costs of LVT proceedings by means such as a costs covenant.

The landlord is then left with three options for paying for an exercise which may be crucial to the management of the building in the interests of all leaseholders:

- He can pay out of his own pocket (generally not an option for RMCs).
- He can pass the cost on to the service charge. This alternative only arises if the leases authorise such matters within service charge expenditure. Even then, service charges are only payable to the extent that they are reasonable, so taking this step may lead to LVT applications by aggrieved leaseholders (particularly if they played no part in the breach) and possibly a shortfall to be met by the landlord.
- If the action resulted from a complaint by another leaseholder (for example, a resident suffering noise nuisance or from a flooded flat caused by a disrepair), the landlord may insist on obtaining a full indemnity and security for his costs from the complainant (most leases allow for such a circumstance, but the landlord's entitlement to an indemnity has been used rarely in the past).

One way or another, it seems likely that innocent leaseholders will end up paying for the breaches of their fellows in many cases. Moreover, many landlords will be more reluctant to use the power of forfeiture. Quite apart from the consequences for RMC's, there could be a significant effect on Right to Manage companies who will sometimes need to seek the landlord's co-operation by taking forfeiture proceedings against leaseholders whose actions undermine their effective management of buildings where RTM has been exercised (see Chapter 7).

Forfeiture for service charges

Forfeiture on the grounds of unpaid service charges has been dealt with separately from forfeiture on other grounds since the inception of section 81 of the Housing Act 1996 (the 1996 Act). This worked as a forerunner of section 168 (see above) by providing in brief terms that no forfeiture proceedings could be commenced unless the service charge had first been admitted or determined by the Court, LVT or arbitration. The most striking difference from section 168 was that a section 146 Notice could be served before any determination.

Section 170 of the 2002 Act introduces two substantial changes to the existing legislation. First, it draws administration charges into the service charge regime. Second, it endeavours to bring that regime into line with the other modifications to forfeiture in section 168. In doing so, it thoroughly amends section 81 of the 1996 Act rather than simply replacing it.

The effect of the amendments may be summarised as follows:

- No section 146 notice may be served nor may any forfeiture proceedings be commenced for non-payment of service charges or administration charges unless the leaseholder concerned has admitted that they are payable, or the LVT, the court or an arbitration has finally determined the amount payable (section 81(1) as amended by section 170(2) of the 2002 Act).
- The only valid arbitration is under a post-dispute arbitration agreement (section 81(5) as amended).
- The right of forfeiture will not accrue until 14 days after any final determination (which includes an appeal or the time within which an appeal may be lodged) (section 81(2) and (3) as amended).

To all intents and purposes therefore, forfeiture for unpaid service charges or administration charges will be dealt with in the same way as any other breach (except non-payment of ground rent), and the general comments above in relation to section 168 can be taken to apply to service charge cases, especially with respect to the costs of proceedings.

Forfeiture of leases not subject to a mortgage

Section 171 of the 2002 Act reserves to the Secretary of State the power to introduce additional requirements restricting forfeiture in respect of premises which are not mortgaged.

Because all forms of forfeiture of residential leases require notice to be given to any third party with an interest in the lease by way of legal or equitable charge (such as a mortgage), there is another link in the chain by which the leaseholder is more likely to be made aware of the proceedings, or at least there will be one party in a position to protect the lease from final termination by forfeiture. That luxury is not available to a leaseholder who holds his property free of mortgage, and an absentee (permanent or temporary) is at far greater risk.

By the time the 2002 Act received Royal Assent, the Government had not come up with any firm ideas for protecting vulnerable leaseholders in such circumstances, so section 171 gives them an opportunity to implement measures quickly when they have been developed.

The role of the Leasehold Valuation Tribunal

By drawing together the provisions of section 168 of the 2002 Act and the amendments to section 81 of the 1996 Act, it is clear to see that for the vast majority of forfeiture cases (except for non-payment of ground rent) the appropriate forum in the first instance will be the LVT.

Under section 168, applications for determinations are to go to the LVT. Meanwhile, by section 27A of the Landlord and Tenant Act 1985 (as inserted by section 155 of the 2002 Act), the LVT has jurisdiction for service charge cases (see Chapter 13). Only forfeitures for unpaid ground rent will go exclusively to the county court, and then only if they satisfy the criteria of section 167 (see above).

The new rules and procedures for LVT cases are discussed at length in Chapter 13, but it is clear that forfeiture issues will greatly increase its workload, quite apart from the other additional jurisdictions which are now allocated to the tribunals.

The alternatives to forfeiture

Of course, forfeiture is not the only weapon available to landlords and property managers; indeed it is not open at all to RTM companies (unless they can persuade their landlords to use it for them). The threat of forfeiture is, however, the traditional 'nuclear weapon' of landlord and tenant litigation. Having said that, forfeiture in the residential leasehold field has been progressively weakened and hamstrung by successive years of legislation, so that it is already used much more sparingly. The provisions of sections 168 to 171 of the 2002 Act will see even fewer attempts to forfeit a lease, or threats to do so.

Landlords of whatever description (freeholders, RMCs, Right to Enfranchise companies, RTM companies) will have to look increasingly at the various alternative methods of enforcement of covenants.

Debt collection

Most enforcement cases relate to non-payment of rents and service charges. There is no restriction on the issue of debt recovery actions in the county court and, if undefended, debt actions can be straightforward and relatively economic. Occasionally, leaseholders' mortgage lenders will accept debt judgments as sufficient evidence that a breach of covenant has been determined to permit them to make payment on behalf of their borrowers, even if the judgment was entered in default of defence.

There are risks with debt recovery cases however, for example:

- A disputed debt claim can be transferred to the LVT to determine the dispute.
- Recoverable costs are strictly limited, particularly in small claims (less than £5,000).
- The debtor may persuade the court to order instalment payments over a long period. With up-to-date charges coming in on top, a clear balance might never be achieved.
- Enforcement of county court judgments can only be carried out in a few ways, and they are not always successful in bringing in the funds. Enforcement can take a long time, and it is not cheap.

Bankruptcy

An alternative to forfeiture which can be just as extreme is to petition for a leaseholder's bankruptcy. This can be done if the debt exceeds £750 and there is no substantive dispute.

Bankruptcy is an expensive process, and is not guaranteed of success in discharging the debt in full. However, because the trustee in bankruptcy will often set out to realise the bankrupt's assets including selling the leasehold property, the landlord can be in a strong position to come to a favourable arrangement with the trustee or any secured creditors such as mortgage lenders.

Bankruptcy has not tended to be a popular option with

landlords, and can be especially distasteful for residents' management companies of which the leaseholders concerned may be members. None the less, the threat of it can be compelling, particularly for the self-employed and company directors.

Injunctions

For breaches of covenant which do not simply revolve around payment of money, it may be appropriate to apply to the court for an injunction (which would not preclude a follow-up action for forfeiture). In cases where the principal objective is to stop the leaseholder (or perhaps a sub-tenant) from doing something, or sometimes when the point is to force the leaseholder to carry out a positive act, a claim for an injunction is likely to achieve at least a short-term result far more quickly than forfeiture.

Examples of breaches of covenant which might lead to injunctions include:

- Unauthorised alterations.
- Nuisance.
- Unauthorised use of the premises.
- Acts likely to render the buildings insurance void.

Injunction applications should generally be supported by damages claims. Because the resultant proceedings require a hearing and usually involve urgent action, it is more common to achieve a higher level of costs recovery (although because the procedure is more expensive, a lower proportionate shortfall in costs can still leave the landlord substantially out of pocket).

Specific performance and damages

Procedurally similar to injunctions, orders for specific performance can be obtained if the primary objective is to require the leaseholder to carry out a positive act to comply with his covenants. The most obviously relevant example would be the repairing covenant. An application for specific performance should be accompanied by a claim for damages for breach of covenant.

Arbitrations or mediation

Reference has been made in this chapter and elsewhere to the fact that the only valid type of arbitration agreement to provide a base for forfeiture or to circumvent the jurisdiction of the LVT is a 'post-dispute arbitration agreement'. It is to be doubted perhaps that very many leaseholders who are simply not settling their arrears (the most common cause of litigation) will co-operate in such an agreement, so they may be a rare occurrence.

While arbitration clauses in leases may be void for the purposes of forfeiture, the 2002 Act does not render them entirely obsolete. Depending on the circumstances of individual cases, it may still be advisable at least to attempt to invoke an arbitration clause simply as a means of dispute resolution.

Similarly, mediation may have a role in a few limited cases. There are specific mediation schemes in some areas, particularly the social housing and retirement sectors.

When the more formal and traditional means of enforcement begin to look so unattractive, a prudent property manager will consider all the various options for dispute resolution which could be used in individual cases.

Checklist for landlords

1. Keep watch on the implementation dates for sections 167 to 171.
2. Review procedures for recovery of small debts.
3. Review procedures for enforcement generally.
4. Look for ways to minimise unrecoverable costs.
5. Consider the strategy for dealing with disputes between leaseholders (particularly in the context of costs).

PART 4

CONCLUSIONS

In the first three parts of this volume, we have looked at the legislative changes contained in the Commonhold and Leasehold Reform Act as they affect the management of blocks of flats, service charges, and dispute resolution and enforcement of covenants. The previous chapters have concentrated on how the law has changed, the resulting procedures, and some of the practical problems which can be anticipated in the context of the various individual headings.

The intention of the remaining chapters is to attempt to draw together an analysis of the overall effects of the 2002 Act, particularly by considering the position likely to be faced by the interested parties in residential leaseholds, and then to examine some of the steps which can be taken to minimise the risks presented by the legislation and to maximise the opportunities which it creates.

Chapter 17

The Effects of the New Legislation

As identified in the introduction to this text, there is a long list of parties who will be affected to greater or lesser degrees by the changes in the Commonhold and Leasehold Reform Act 2002. To repeat the most obvious categories:

- Developers.
- Institutional landlords.
- Local authorities and Housing Associations.
- Leaseholders.
- Property managers.
- Surveyors.
- Valuers.
- Arbitrators.
- Solicitors and other lawyers.
- Residents' Management Companies.
- Right to Enfranchise companies.
- Right to Manage companies.

This chapter will concentrate on those parties who will feel the effects most directly.

Right to Manage companies

Right to Manage companies are created by the 2002 Act. The whole concept of RTM and the rights and duties of RTM companies are examined in detail in Chapters 1 to 8 of this volume. What has tended to be forgotten about RTM companies in the build-up to the 2002 Act is that they will be faced with most of the other provisions in the Act which increase the demands upon managers of leasehold property. RTM companies are as subject to the new service charge requirements, for example, as any landlords.

Leaseholders intending to exercise the Right to Manage must bear in mind (as should their advisers) that taking over management will mean assuming a broad range of statutory responsibilities, such as:

- The taxing new requirements for consultation on major works and long term agreements (see Chapter 11). These obligations will pose particular problems immediately from the acquisition of RTM, and will supply an incentive to stay with the outgoing manager's existing contracts as far as possible (which may go against the grain for RTM company members). RTM companies will depend upon full recovery of their expenditure through the service charge, and they cannot afford to take the risk of major reductions caused by non-compliance with the consultation requirements.
- RTM companies will not have any slack when it comes to the exercise of the administration side of their management functions. They will have no funds to carry this work for free; after all, the company members will already be responsible for the company's set-up and running costs which cannot be put to the service charge. Consequently, they will need to recover what they can by way of administration charges, which are now subject to the scrutiny of the LVT (see Chapter 10).
- RTM companies will be subject to the vast majority of the new accounting requirements for service charges contained in the 2002 Act (see Chapter 12). For example: the more detailed statements of account; the need to send out summaries of leaseholders' rights and obligations; certification; leaseholders' rights to inspect documents; the duties of trustees of service charge funds. Similarly, RTM companies will be bound by the amended rules for insurance information (see Chapter 14).
- RTM companies will be hampered by the new right for leaseholders to withhold service charge payments in certain circumstances (see Chapter 12).
- RTM company directors may be held personally to have committed an offence in the event of non-compliance with accounting requirements.
- Generally, RTM companies may find themselves facing LVT applications from leaseholders or landlords on a whole range of issues (see Chapters 12 and 13).
- When it comes to enforcing leaseholders' covenants (especially payment of service charges, the life-blood of every RTM company), RTM companies seeking to persuade their landlords to take or threaten forfeiture proceedings will find that the new restrictions on forfeiture (see Chapter 16) will mean that the weapon is not what it was. Consequently, it may be necessary

to look at the less effective alternatives to forfeiture, which are likely to have a delaying impact on service charge collection.

Right to Enfranchise companies

Right to Enfranchise (RTE) companies will be in generally the same position as RTM companies, with the obvious exception that they will also own their freeholds. Indeed, it may transpire that many RTE companies are converted from RTM companies. RTE companies are also potentially subject to claims for the Right to Manage in future years.

The position of RTE companies is looked at in greater depth in Chapter 9.

Residents' Management Companies

Whereas RTE companies will be of a uniform structure as a creature of the 2002 Act, Residents' Management Companies (RMCs) come in all shapes and sizes. They may have acquired their freeholds by enfranchisement or by negotiation, they may have been put in place as part of an original development scheme, or they may simply have a role as manager rather than freeholder. Many RMCs have been in existence for several years. As they have not been created or fundamentally altered by the 2002 Act, RMCs will have had less reason to take notice of the Act's provisions, especially if they do not employ professional managing agents.

Consequently, RMCs are probably at greatest risk from the new legislation.

Precisely what effects will be felt by individual RMC's will depend upon their status at the property concerned: in other words, whether they own the freehold, whether they hold a head-lease, whether they are manager parties to the leases, or whether they have no formal recognition within the freehold or leasehold titles. In general terms, RMCs will have to deal with all the same issues as RTM companies and (if they own their freehold) RTE companies (see above).

RMC directors, who may have been in place for some years, will have to learn entirely new rules and procedures for such matters as:

- Consultation.
- Accounting.
- Administration charges.

- Demanding ground rent, service charge and other monies.
- Enforcement.

The changes to the statutory requirements are so far-reaching and complex in many cases that RMCs who manage their blocks without employing professional managing agents may need to re-visit the policy regarding out-sourcing management, or at the very least take advantage of any opportunities for training and education for directors.

Developers

Apart from the Commonhold elements of the 2002 Act (which are not the subject-matter of this text), the Act has little direct impact upon developers, unless they intend to retain the freehold (see under 'freeholders' below). There may be substantial indirect impact however.

Developers of leasehold or mixed use estates frequently have to take into account when drafting leases the needs of their successors to the freeholds (under option agreements for example) or of the residents' management companies whose role will crystallise when the development is complete. Developers and their solicitors may find that they are requested to make subtle changes to their leases to allow for the requirements of the new legislation.

Modifications may also be needed to such things as managing agents' contracts which are entered into before the completion of the development because of the new provisions on consultation. Contractors may be looking for shorter or longer terms for example, depending on the circumstances. Similarly, services like lift maintenance or security systems may need to be addressed.

Freeholders

Leaving aside resident-owned freehold companies, the majority of remaining freeholders are investors of one sort or another. Their principal concerns will be economic therefore, both in terms of preserving the value of their assets in the long term and in maximising the income from rents, fees and commissions.

Obviously the elements of the 2002 Act which make enfranchisement more straightforward, and fix marriage values at what freeholders will see as more favourable levels for leaseholders, will have some negative impact on the value of freehold reversions. More subtle effects will be felt on both capital and income as a result of some of the leasehold reform elements of the Act. Examples include:

- Right to Manage cases will have a direct effect on income, but may also have a detrimental effect on the value of the reversion if the RTM company fails properly to perform the landlord's covenants.
- The greater facility for leaseholders to obtain information concerning service charge and insurance matters will lead to greater transparency, so that some landlords will feel a squeeze on items such as commissions and referral fees.
- The wider consultation requirements on long term agreements may distance some landlords from their usual or retained contractors.
- All landlords will notice an increase in administration costs, not all of which may be recoverable.
- The difficulties in enforcement will delay cash flow, and the lower levels of recoverable costs will make enforcement considerably more expensive.

Social landlords

The whole topic of how the legislation affects local authorities, housing associations, charitable housing trusts and other social landlords deserves a book of its own. There is not the space in this volume to cover the ways in which social landlords will have to review their procedures and general strategy, especially as different types of landlord are subject to different effects. For example, the Right to Manage can be exercised against a housing association but not a local housing authority (see Chapter 1); there are different rules for consultation depending upon whether public notice is required to be served.

Generally, the drift of the legislation is to bring the public sector into line with the private sector as far as is considered practicable at this stage.

Managers of social housing should be taking strenuous steps to ensure they have appropriate advice and training on hand.

Property professionals

Any professional working in the field of residential leasehold property is likely to have spent a considerable amount of time coming to terms with the leasehold reform proposals over the last few years. Often they will have been concentrating on the potential impact upon their clients. It is too easy to overlook the consequences of new legislation on a professional's own business.

Surveyors (who are not also property managers) and valuers are likely to find that their instructions come more from leaseholders' groups and companies and less from large corporate landlords as the Right to Manage and the Right to Enfranchise start to bite. They may thus encounter more pressure on fees than previously. Surveyors in particular may notice an increase in appointments to report to leaseholders on existing landlords' work programmes, service charges and management generally.

Arbitrators (who are usually surveyors in a landlord and tenant context) could see a drop in referrals as arbitration clauses in leases are rendered largely obsolete. There may be an increase however in the number of surveyors and valuers appointed to the Leasehold Valuation Tribunal as it endeavours to fulfil its new jurisdictions.

Solicitors are likely to find themselves called upon to advise and assist leaseholders' companies more than before in terms of exercising the Right to Manage or dealing generally with all the new complexities in the service charge and other areas. Forfeiture proceedings will decline markedly, but it remains to be seen whether landlords and managers instruct solicitors and barristers for the greatly increased number of LVT cases which can be expected. Alternatively, there could be a sharp rise in debt recovery actions in the county court.

Clearly, managing agents will be most directly affected, since the changes involve practically every aspect of their work. Some of the challenges which agents will have to meet include:

- Agents could be on the receiving end of RTM, or they may be assisting an RTM company to exercise the right. If the former, they could be in a difficult position: possibly helping the landlord to stave off the claim, or at least easing the transition in the interests of their existing client; meanwhile, they may wish to distance themselves from the landlord in order to encourage the RTM company to continue with their services. If the latter, agents may well find themselves doing a good deal of development work free of charge for no end result. In the minimum four month period it takes to acquire RTM, there is a risk that it will be abandoned or ruled invalid. Even if the claim to RTM is successful, there is no guarantee that the agents will be reimbursed. Agents will have to consider charging separately for preparing a claim to RTM, or reserving the right to charge if the claim comes to nothing.

- Managing agents' agreements will now be subject to consultation with all leaseholders at a block under section 20 of the 1985 Act if they exceed one year and any one leaseholder will be charged over £100 per annum in respect of their fees (see Chapter 11). Many agents' charges are in excess of £100 per flat per annum (especially in London and the South East), and terms of three or five years are common. Leaseholders can make their own nominations and landlords must present at least two estimates for consideration. Inevitably, agents' fees will have to be competitive, and some will have to improve their customer relations skills.
- Some agents have been accustomed to supplementing their fees through a variety of administration charges. Such charges will now be subject to the LVT in much the same way as service charges. The onus will be on the agent to demonstrate that they are reasonable.
- Across the board, property managers' workloads will increase as a result of the Act. The new consultation procedures and accounting requirements alone will create substantial additional levels of administration and overheads. The Government has recognised that the cost of the new rights and protections for leaseholders will be an increase in management charges. At the same time however, agents' fees and charges will be scrutinised more than ever before. This places agents in a difficult position. If they estimate too high they may lose business opportunities; if they quote too low they might put themselves under real financial pressure.

It can be anticipated that there will be great changes among managing agents when the consequences of the 2002 Act work through the system. Smaller firms may depart from the scene while larger enterprises look for economies of scale. Meanwhile, the drive among many agents to provide a better educated and more professional service (as evidenced by the establishment of the Institute of Residential Property Management (IRPM)) is an encouraging sign that managers are rising to the challenge (the subject of the final chapter).

Chapter 18

Preparing for Change

The Commonhold and Leasehold Reform Act 2002 brings in a vast range of challenges for all involved in residential leasehold property, particularly those responsible for managing blocks of flats. As with all new legislation of any substance, new threats and opportunities are posed. This concluding chapter will look at some of the possible ways of protecting against the threats and grasping the opportunities.

Preparing for change

Preparation of course is all about anticipation. In order to anticipate, there must be access to information. Consequently, the starting points are education and training.

Professionals in the property field have a duty to themselves and their clients to ensure they take advantage of every available opportunity for training. Among others, the Association of Residential Managing Agents (ARMA) and the Association of Retirement Housing Managers (ARHM) have been playing major roles in providing education for property managers in relation to the 2002 Act, and will continue to do so. The Leasehold Advisory Service (LEASE) provides very useful publications for professionals and lay people alike.

Deserving particular attention is the Institute of Residential Property Management (IRPM), whose birth has coincided with the arrival of the 2002 Act.

Not all those responsible for managing blocks of flats are professional managing agents. Many buildings are run independently by leaseholders, usually by directors of RMCs, but now increasingly by other organisations created out of the 2002 Act: RTM and RTE companies.

Inevitably, lay managers will have much more restricted access to education and training. Perhaps unfortunately, RMC's and RTM and RTE companies are not exempt in any way from the effects of the new legislation; indeed, in many ways such companies are likely to be hit hardest by the strict new regimes on such issues as consultation, service charge accounting and enforcement. This is where LEASE's role in publishing excellent guidance notes is most valuable.

Reading in isolation only helps to a limited extent however. Conferences, seminars and training workshops will often clarify matters and provide useful opportunities for discussion and trying out ideas. In the case of legislative changes which have far-reaching practical consequences, practical training sessions can be most beneficial.

Experienced professional property managers have a potential part to play in the education process, whether they come from the ranks of RICS, ARMA, ARHM, CIH, IRPM or other organisations. As they will be the best attuned to the legislation, old and new, they will also be best placed to pass on their knowledge and expertise. It is probable that increasing numbers of leaseholders will be assuming the management of their blocks in one way or another. There will always be a role for property managers, but it may be different from before. RMCs or RTM or RTE companies are likely to need professional help from time to time, but they may refer to professionals only occasionally, perhaps for specialist help in certain areas of management. To fill the gap, it could be that property professionals would do well to offer a more diverse service, with consultancy and training featuring prominently.

The areas where training will most probably be required either directly or indirectly as a result of the 2002 Act include:

- RTM: exercising the right.
- RTM: managing the block after acquisition.
- RMC and RTE: setting up and running the companies in accordance with the statutory requirements.
- RMCs, RTM and RTE companies: advising on potential corporate and personal liabilities and obtaining appropriate insurance cover in all its forms.
- Administration charges and improvements: the new regime.
- Consultation on works and long term agreements: the new statutory requirements and procedures.
- Service charge accounting: the new statutory requirements and procedures.
- The risks of non-compliance with the new service charge regime.
- Service charge recoveries in the light of the restrictions on forfeiture and the role of the LVT.

- Insurance: the new statutory requirements and procedures.
- The extended jurisdictions of the LVT and its new procedures.
- How to conduct a case before the LVT.
- The changing rules on ground rent.
- Enforcement of covenants generally.

Also, with newer management companies, especially RTM companies perhaps, essential skills to be learned will include budgeting for expenditure and long term maintenance programmes.

Training is only one step of the way however. Constant reviews of procedures will need to be instigated, from regular checks on the implementation of the legislation, through keeping up-to-date on its interpretation by the courts and tribunals, to more mundane items such as ensuring that the wide range of prescribed forms are updated as appropriate.

Rising to the challenge

Leaseholders

The 2002 Act represents a further major overhaul of the law relating to residential leasehold. It will not be the last. Those actively involved in leasehold property face a constantly shifting tide of legislation and case law. Meanwhile, developers are continuing to build blocks of flats in great numbers, and the days of mixed use developments are beginning to return, often on a very large scale. New developments bring new leases, hopefully more user friendly than before.

Each new development scheme and each change in the law brings a new challenge for property managers, of whatever description.

From the point of view of leaseholders, the 2002 Act introduces a host of new rights, but many of them also usher in new responsibilities. The RTM is a huge step forward for flat-owners, empowering them to have the major say in the management of their properties without having to go to a Court or tribunal to prove a case against their landlords or existing managers. RTM needs to be taken very seriously though if it is to work in practice. RTM is only one means to an end, and anyone considering using it should be thoroughly prepared, forewarned as to the obligations it imposes, and always bearing in mind the alternatives.

The other reforming elements of the Act will give leaseholders opportunities for gaining more influence over the management of their properties in any event. The increased powers of scrutiny and the right to withhold payment of service charges in appropriate cases may achieve similar end results to exercising the Right to Manage, without the risks or responsibilities.

All of the reforms, including RTM, will work best if leaseholders act together to the maximum possible extent. The challenge for leaseholders is to use their rights responsibly and collectively. If they do, the 2002 Act can present the chance of a massive improvement for leaseholders.

Property managers

For those who earn their livings managing leasehold property, and who intend to carry on doing so, the Act is likely to lead to changes in approach and marketing strategies. Particularly the RTM and the new requirements for consultation on long term agreements will mean that managing agents can no longer rely on holding on to appointments without challenge just because they are put in place by developers or freeholders. On the other hand, the variety of new leaseholders' companies who can take over management will provide a greatly expanded market for managers.

Property managers looking to succeed in the new climate will need to refine some or all of the following skills in order to retain business and develop new sources of work:

- Communication. One of the most consistent causes of criticism of managing agents recorded in LVT cases is that they do not communicate adequately or at all with leaseholders. Although many managers have been improving in this regard, there is always more that can be done. Managing agents who have worked consistently with residents' management companies may have a headstart, as they will be experienced in knowing when and when not to make contact with directors. Quality rather than quantity is often the key.
- Knowledge. Managing agents are required by the nature of their positions to have a good working knowledge of a wide field of law and practice. For example, they need to be familiar with at least the basics of landlord and tenant, company, employment, health and safety and land law. With a more diverse market, they will now have to be able to demonstrate that capability with paper qualifications as well, which highlights the relevance of the likes of IRPM.
- It is essential that managers can show that they have good

working relations with other contractors and suppliers (including the likes of insurers, surveyors, accountants and lawyers). Clients will want the reassurance that their managers can bring in the appropriate assistance when it is needed, and that the source can be relied upon to have the requisite skills and expertise.

- Meanwhile, agents who can provide from their own resources comprehensive services in allied disciplines will significantly advance their own marketability. By supplying a package, agents can keep costs down overall, while maximising their fee income. Many managing agents are perfectly capable of offering services such as debt recovery, advocacy (at LVT hearings), company secretarial and arranging insurance; they may also gain valuable fees by providing ancillary services including letting and management of individual flats.
- Managing agents will have to show a professional approach combined with demonstrable qualifications to attract entirely new business. Potential clients will have to start their search for a manager somewhere, and qualifications are likely to be of help in establishing a shortlist.
- Connected to the previous point, agents would benefit from membership of relevant organisations (ARMA, ARHM, RICS and so forth). Many new clients will be looking for visible complaints procedures and evidence that agents comply with an appropriate code of practice.
- Similarly, managers must be able to demonstrate adequate professional indemnity insurance cover.
- For new business, and for retaining existing business to a certain extent, managing agents will benefit from being able to show that they are genuinely independent. Agents have been perceived historically by some leaseholders as tools of the freeholders. This is partly inevitable, because the larger landlords have tended to use the same agents across their property portfolios, and indeed many agents have been created out of the landlord companies. In reality, this is far less true than it used to be, but while the perception persists, agents need to guard against it.
- Finally, on a practical level, managers will have to show a willingness to attend meetings out of office hours and such matters, but laying down reasonable limits and encouraging discipline on the part of management company directors will not generally damage marketability so long as the policy is established from the outset.

The other principal challenge posed by the Act concerns the way managers carry out their tasks. They will have to confront the need for new and improved procedures for matters such as:

- Consultation procedures under section 20 of the 1985 Act.
- Accounting for service charges and banking.
- Enforcement of covenants, especially debt recovery.
- Managing agents will have to refine the ways in which they justify their fees. Overheads will increase as a result of the new measures in the 2002 Act, but at the same time the new market conditions will mean that fees must be competitive. That does not mean inevitably that fees will have to be reduced, but value for money must be demonstrable. Time recording and transparent calculation methods will assist agents both in negotiating fee agreements and in fending off challenges to the LVT by being able to show that they are reasonable. In the meantime, managing agents will have to look to ways of keeping overheads to a minimum by a proper use of systems and delegation of tasks to members of staff at appropriate levels of qualification and experience. The bottom line is that agents are entitled to make a reasonable profit. If they cannot do so, the services they provide will not exist.

Freeholders

For freeholders (other than RMCs and RTE companies, which are discussed elsewhere), the challenge arising out of the 2002 Act is how to continue to derive a reasonable level of profit from their capital investments. Clearly, the major purpose of the 2002 Act and its legislative predecessors is to redress the balance of power between freeholders and leaseholders. Accordingly, freeholders can be expected to lose out in a number of ways: the reductions in marriage value; the greater ease of enfranchisement; the introduction of the Right to Manage; checks on administration charges; increased controls over service charges; the new rules on ground rent and the landlord's nominated insurer; the additional restrictions on forfeiture. Each of these reforms will have a negative effect on freeholders' incomes, directly or indirectly.

On the other side of the equation, there is little which freeholders can do to advance their position simply within the boundaries of the 2002 Act, save for the obvious step of preserving their position by enhancing the value (and thus the valuation for enfranchisement purposes) of their properties and improving their relations with leaseholders so as to remove incentives for leaseholders to take over their blocks.

It is foreseeable that there will be some contraction among freeholders leading to economies of scale. The 2002 Act does not completely preclude profit being made in the leasehold sector by successful freehold investors.

Minimising the risks

To a large extent, the risks posed by the 2002 Act can be minimised by the same methods as are suggested above to rise to its challenges. The Act does carry perils (which are discussed at length in the previous chapters), principally for landlords and property managers, but also for leaseholders who own or manage their own properties and those who are considering doing so.

It may be helpful to summarise in this context the measures which can be taken by pretty well everyone concerned to keep the risks to a minimum:

- Preparation. The Act imposes severe sanctions for non-compliance with its requirements. Some of the individual elements pose dangerous traps for the unwary (particularly the new innovations like RTM). Ignorance of the law is no excuse, so steps must be taken to ensure all concerned are fully conversant with its provisions.
- Training and education. The new law and the new skills required to comply with it should motivate practitioners and lay people alike to grasp available opportunities for education. The professional organisations such as ARMA, ARHM, RICS and IRPM have a heavy responsibility here, a responsibility which they have already shown a willingness to accept. Training must allow for continuing development, as updates on interpretation of the Act and how it is working in practice will constantly be needed.
- Professional advice and assistance. Because of the complexities introduced by the 2002 Act, and in the interests of good practice in any event, it is likely that leaseholder organisations of various types will recognise a greater need for professional advisers and managers. Where it is needed, prudent leaseholders and their companies will search out professionals who are adequately insured, trained and

qualified. In the case of managing agents, this represents an excellent opportunity for those who have taken the trouble to join relevant organisations and take up available qualifications. For ancillary service providers such as lawyers and accountants, there are still relatively few who practise consistently in the leasehold field (and thus, opportunities are likely to abound for those who do). Potential clients should check for relevant expertise when approaching advisers.

- Systems. The 2002 Act will lead inevitably to an increase in administration and bureaucracy, particularly in areas like consultation and service charge accounting. Relatively new to this field, there are a host of prescribed forms and procedures. A good deal of long-term expense can be saved by the short-term investment in systems.
- Communication and co-operation. Historically, the landlord and tenant field has been invested with considerable mistrust and ill-will between the parties. The Act gives a substantial incentive to freeholders, leaseholders and managers to work together. Greater transparency and better listening skills will aid an improved understanding of each other's expectations and difficulties, and thus head off many intractable disputes. This is perhaps especially important for landlords, because disputes are likely to lead to loss of management and income.
- Insurance. The need for insurance against personal liability in terms of directors' and officers' cover, employers' liability, public liability, and professional indemnity will be greater than ever, particularly as leaseholders begin to assume management for their buildings in ever increasing numbers.
- Funding. The need to arrange finance will be particularly important for RTM companies who will need to fund their management functions immediately upon acquiring the right. Other landlords may find that they need to look at this eventuality as well, especially other forms of leaseholder controlled management company, in the light of the new rights for leaseholders to challenge and delay service charge payments.

Conclusion

The law relating to blocks of flats has always been a minefield. The 2002 Act attempts to show leaseholders a path through it, but it does not set out to clear the mines completely. Indeed, some more are set on the way.

Undoubtedly, the Act contains many tangible reforms which will prove advantageous to leaseholders in the fullness of time: the Right to Manage; easier enfranchisement; greater protection of service charge funds; increased rights of consultation and scrutiny; further restraints on forfeiture. Very few of the Act's provisions will be straightforward in their implementation however, and it is difficult to see at this stage how well they work in practice.

For leaseholders, landlords and practitioners alike, it is essential that the Act is made to work and that its more problematic effects are avoided. It is hoped that this book will be of assistance in those endeavours.

Glossary of Terms

ARMA	Association of Residential Managing Agents
ARHM	Association of Retirement Housing Managers
CIH	Chartered Institute of Housing
CLRA	Commonhold and Leasehold Reform Act 2002
LEASE	Leasehold Advisory Service
LVT	Leasehold Valuation Tribunal
RICS	Royal Institution of Chartered Surveyors
RMC	Residents' Management Company
RTE	Right to Enfranchise
RTM	Right to Manage

Useful Addresses

Association of Residential Managing Agents, 178 Battersea Park Road, London SW11 4ND. Tel: 020 7978 2607. www.arma.org.uk

Association of Retirement Housing Managers, 3rd Floor, 89 Albert Embankment, London SE1 7TP. Tel: 020 7820 1839. www.arhm.org

Institute of Residential Property Management, 178 Battersea Park Road, London SW11 4ND. Tel: 020 7622 5092. www.irpm.org.uk

Leasehold Advisory Service, 70-74 City Road, London EC1Y 2BJ. Tel: 020 7490 9580 or 0845 345 1993. www.lease-advice.org

Leasehold Valuation Tribunal, Residential Property Tribunal Service, 10 Alfred Place, London WC1E 7LR. Tel: 020 7446 7700. www.rpts.gov.uk

Royal Institution of Chartered Surveyors, 12 Great George Street, Parliament Square, SW1P 3AD. Tel: 0870 333 1600. www.rics.org

Index